'*The Missing Musk* is not really about finding answers, but challenging the systems that govern our thinking . . . we are asked to celebrate the "confusions and contradictions" of nature, acknowledging what we cannot know while valuing all perspectives, from the metaphysical to the biological'

Times Literary Supplement

'A brilliantly researched investigation into some of nature's most enduring mysteries. Never less than utterly fascinating and intriguing'

Neil Ansell, author of *The Last Wilderness*

'Thought-provoking and enlightening, *The Missing Musk* is a perfect mix of whodunnit and wonder for the natural world. A truly unique, fun and inspirational book that dispels common myths without detracting from the wonder of its subject'

James Aldred, author of *Goshawk Summer*

'A joyful celebration of intelligent curiosity. Absorbing, fascinating, and a positive invitation to explore the natural world around us'

Lev Parikian, author of *Into the Tangled Bank*

'Impeccably researched and endlessly fascinating, *The Missing Musk* is a captivating dive into the mysteries of nature'

Lee Schofield, author of *Wild Fell*

Bob Gilbert is the author of *Ghost Trees*, which was long-listed for the Wainwright Prize and Folio Prize, and *The Green London Way*. He has written a newspaper column on urban wildlife for the last twenty-five years and is a contributor to TV and radio, including BBC Radio 4's 'The Susurrations of Trees' and 'The Passion in Plants'. Bob has also been a long-standing campaigner for inner city conservation and is patron of The Garden Classroom, a charity that promotes environmental education in London. He lives in London with his family, Ash the greyhound, and a peripatetic tortoise.

Also by Bob Gilbert

The Green London Way
Ghost Trees

The Missing Musk

A Casebook of Mysteries
from the Natural World

BOB GILBERT

sceptre

First published in Great Britain in 2023 by Sceptre
An imprint of Hodder & Stoughton Limited
An Hachette UK company

This paperback edition published in 2024

1

Copyright © Bob Gilbert 2023

Internal illustrations © Tom Hodges-Gilbert 2022

A CIP catalogue record for this title is available from the British Library

Paperback ISBN 9781529356007
ebook ISBN 9781529355987

Typeset in Sabon MT by Hewer Text UK Ltd, Edinburgh
Printed and bound in Great Britain by Clays Ltd, Elcograf S.p.A.

Hodder & Stoughton policy is to use papers that are natural, renewable
and recyclable products and made from wood grown in sustainable
forests. The logging and manufacturing processes are expected to
conform to the environmental regulations of the country of origin.

Hodder & Stoughton Limited
Carmelite House
50 Victoria Embankment
London EC4Y 0DZ

www.sceptrebooks.co.uk

To my sister Gill.

Not least for telling me to get on with this book.

Contents

The way of this mystery, the wonder of its process, is not justified by its endpoint. It wanders ahead in time and in space by no terribly linear path. Yet each step matters. The mystery draws us onward. We are always trying to figure it out; to discern our way; to gather clues, hints, and signs . . . Along the way we solve one problem after the next. But the content that concerns us here may pose a real enigma: when we think we have finally got it, have we already lost it?

from *On the Mystery*
Catherine Keller

The universe is not only stranger than we imagine, it is stranger than we can imagine.

Albert Einstein

Introduction

In 1913 all the musk plants in the world stopped smelling. Here was a plant that was hugely popular, grown everywhere from country gardens to working-class windowsills, and entirely for the fragrance that was reflected in its name. And yet that fragrance had, suddenly and globally, gone. When I first heard the story, told me by an elderly man as we sat round his garden table in the Yorkshire Dales, I was understandably sceptical, even though he seemed to me old enough to have been there at the time. But he swore to the truth of his tale, though he could provide no explanation for it. In fact, it took only a little subsequent digging for me to substantiate the gist of what he had said. It was a plant that had once been the centre of a massive horticultural trade but it existed, in this country now, only as a sporadic wetland wild flower; and in all of my plant books the descriptions of its once vaunted fragrance were prefixed with the puzzling word 'former'.

The juxtaposition of mystery and natural history was something I could not resist. Even as a child, brought up in a Christian fundamentalist household in south London, I had often escaped from the burden of didactic sermons and incomprehensible injunctions by wondering at the origin of the silky egg sacs that hung in a garden shed, the inexplicable comings and goings of green bottles on a park leaf, or the hunting habits of the wolf spiders that scuttered across our prefab path. As for the musk plant, I already knew it a little. I had encountered its prettily shaped butter-yellow flowers now and then, in a damp corner of a Suffolk campsite or alongside the River Kelvin in urban Glasgow, but this part of its story was new to me. It unlocked my inner

Sherlock and I longed to don the metaphorical deerstalker and investigate. But then life intervened, as life often does, with work and children and even the washing up. I moved on to other things and it was only after many years, and prompted by another encounter with the plant in Glasgow, that I actually got round to some serious research. What I uncovered was a story involving some of the most significant scientific figures of the day. And a story, too, of bitter rifts in academic institutions, of eccentric theories and competing ideas, and of a cantankerous plant collector who had himself met a mysterious, and probably murderous, end.

But this is to race ahead and those intervening years had been serving a purpose of their own. Alerted by the mystery of the musk's missing musk, I had begun to look out for other such enigmas, and what you are alert for almost always appears. I discounted stories such as the Loch Ness monster, the Beast of Bodmin and the Devonshire Black Dog, and any of those other hyper-curiosities that lurk in the imaginative shadows. What I was most interested in was the established accounts of everyday plants and animals that nonetheless contained an element of the unexpected and the so-far unexplained. And so I began to accumulate 'natural mysteries'.

In 1972 the people of the Dorset town of Blandford Forum were afflicted in their hundreds by serious bites and lesions to their limbs, related to the sudden appearance of a mysterious black fly. When it happened again in 1988, stories appeared in the national press, questions were asked in parliament and rumours ran locally rife, among them the suggestion of nefarious activities in a nearby military camp. A locally brewed beer had even been named 'The Black Fly' in its honour. A much more long-running, and far more widespread story involved the appearances of an unexplained substance known as 'star jelly'. Gelatinous globs, transparent, white or even purple, they could crop up anywhere, sometimes in a single lump, sometimes in quantity and over a wide area. Known about for hundreds of years, they had once been associated with meteorite showers. More recent explanations remained both divisive and doubtful and the substance had entered popular culture in a particularly sinister form. Also aspiring to the astral plane was the story of the tardigrade. This tiny but abundant creature, fondly and familiarly known as the 'moss piglet' or 'water bear', has the ability to survive the most extreme conditions, beyond any, in fact, to be found on our planet. It has fascinated writers of both science and science fiction and been championed by evolutionists and creationists alike. Could its strange qualities really be evidence, as some serious scientists had suggested, that its origin is extra-terrestrial?

Other stories lay much closer to my own front door. My local cemetery had a population of hedgehogs, probably the only surviving colony in the borough. I had read of the hedgehog's strange habit of 'self-anointing', of regularly slavering itself and covering its spines with saliva. No satisfactory scientific explanation for this extraordinary behaviour has yet been supplied. Even closer to home were the stories I was hearing of a population of mosquitoes living an entirely subterranean life on the London Underground. Was it true, or was it an urban myth? And, if true, how on earth, or in this case beneath it, had they come to be there and by what evolutionary changes had they managed to survive?

The musk was not to be the only botanical entry in this growing portfolio. I learnt, for example, of something called the 'great oak change', a suggestion that sometime in the twentieth century this iconic tree had completely, and inexplicably, changed its approach to propagation. And then there was the yew, that dark and brooding tree whose very presence seemed to encapsulate mystery. What was the origin of its ancient association with the churchyard, where three quarters of all our oldest and largest specimens are to be found? The relationship has spawned speculation across centuries with competing theories including the biological, the historical and the downright mystical. At the other end of the scale was the tiny lichen, that seemingly simple life form that drapes itself in scaly bunches on woodland trees or spreads as dry crusts on city walls. One of the earliest forms of life on dry land, this apparently primitive 'plant' confused and divided generations of scientists who, in their differences over its identity, descended into vituperative debate and sometimes slander; perhaps not surprisingly so, for the question of whether a lichen is one organism or many turns out to have profound implications on the way we see the world as a whole.

Another group of stories seemed, in their substance, as much sociological as they were scientific. How, for example, had one of the most shameful episodes in British military history led to the arrival in this country of one of our commonest roadside plants? And what was the truth behind the reputed connection between Father Christmas and the fly agaric, that common and hallucinatory mushroom that populates both our woodlands and the pages of our children's picture books? What, too, of the story of the bulrush? Sometime in the mid-nineteenth century the name had been stolen from one common waterside plant and applied, universally, and incorrectly, to another. It was an identity crisis apparently caused by its appearance in a popular Victorian painting, yet the true identity of the painting, and even of the painter, remains uncertain, varying according to the account you hear. It seemed another sort of riddle, and one worth unravelling. And on the question of identity, what of the problem of the Duke of Burgundy, or rather of the small brown butterfly that was inexplicably named after him? I have glimpsed it only occasionally, on the steep grassy slopes of Rodborough Common in Gloucestershire, or on the high downs behind Gilbert White's house in Hampshire. Though originally bestowed with a variety of awkward appellations, such as 'Mrs. Vernon's Small Fritillary', the consensus had gradually settled on the French Duke, yet not a single lepidopteran authority could find any reason, physical, historical or otherwise, for this connection with an obscure member of an overseas aristocracy. Here, surely, was another enigma demanding to be examined.

* * *

Such was my collection of 'cases', and what had started as a matter of mere curiosity was now beginning to stake a rather larger claim. On scruffy bits of paper and in dog-eared pocket notebooks, I was collecting 'clues', amassing notes and references and, as the process became increasingly unruly, trying to impose some order by organising them into files. To further aid my progress, I even ordered a crate of the specially brewed Black Fly ale, though half of it still sits unopened in my study. Perhaps the idea that all this might one day make up a book had always been at the back of my mind, though, in fact, a couple of others were to get in the way first. I thought of it as my 'nature detective's casebook' though I kept that name to myself, nervous that it echoed too much of an earlier and more innocent era of 'Walks in the Wildwood', 'Adventures with Nature' or '301 Things a Bright Boy Can Do'. It was only after the publication of *Ghost Trees* that the time, and the opportunity, seemed finally to have arrived – much to the relief of those friends and family who had heard me interminably expound the idea.

It was not easy to reduce those collected files to a favoured and manageable few. I had already rejected, as beyond my remit, accounts of unexpected species said to survive in surprising spaces; the wallabies that had lived for so long on the Staffordshire Roaches, the freshwater jellyfish that flourished in a Sheffield canal, the stories of scorpions living and breeding in my own East End. They had each been set aside, even though I would have dearly liked the excuse to look for them all. Most of the more 'sociological' stories were the next to go. It was not that they were any the less interesting. My early training had been in sociology. It had changed the way I saw the world and even shaped my approach to natural history, and I owed it a great deal. But these stories seemed to belong to a slightly different genre and there was always the sneaking suspicion that one day they might make up a book of their own.

Personal experience played its part in determining my selection. It was one of my own children who had, as a schoolboy, first introduced me to tardigrades, whilst I had encountered the gelatinous star jelly on a family holiday. I had, too, long had a love affair with the colours of lichen, especially the orange, rust-brown, lemon-yellow and lime-green crusts encountered on gravestone, mountain or seaside rock. I had also an existing, if rather more ambivalent, relationship with the yew. I once made a radio programme on a subject we had called 'susurrations', the sounds made by the leaves of trees in the wind, and whether you could, as Thomas Hardy had suggested, use them to distinguish the different species. We had recorded the papery rustling of the oak tree, the more metallic murmuring of beech, the soft conversations of pine trees and the whispering sibilance of birch. We had ended by standing beneath, and virtually within, an ancient yew tree, and, here, instead of sound, we recorded silence; the deep, enwrapping silence of the yew, a tree that seems not so much to emit sound as to absorb it.

All these subjects, then, made it to the final cut. Beyond that I wanted to include, as far as possible, subjects that were accessible to all. Much of my previous work, whether walks or talks or writing, has been about the extraordinary in the everyday, about paying attention to the abundance that exists all about us and even in the most unpromising places; the pearlwort that grows in pavement cracks, the yellow heads of cat's-ear in an uncut lawn or the spiders that spin their untidy webs across the dim lights of a pedestrian underpass. My choice, I hoped, would continue to respect that approach. The most startling ghost is the one that appears behind the kitchen door and it was not the exotic I was after; it was the mystery that is heightened precisely because it is embedded in the ordinary.

Set out in these chapters are my final six choices; the enigmas I set out to explore and the adventures they involved me in. I studied in the library of the London Natural History Museum under the continual, stony but, I hoped, benevolent gaze of the bust of Charles Darwin. I sat in the galleried halls of the Gladstone Library where all that great man's own books are housed. I ranged rows of shelves at the Wellcome Collection, where the book I most needed always seemed to require the help of the tall library ladder or reduce me to my knees on the floor. The British Library became a second home, the extravagant height of its atrium somehow seeming to combine the atmosphere of a major railway terminus with that of a modern cathedral. I held handwritten pages from the journals of explorers in the chilled basement rooms of the Royal Horticultural Society and inspected sheets of dried specimens in the herbarium at Kew, where the high, metal-galleried halls seemed a fair copy of those in Pentonville Prison. It was the airy reading room of the adjacent Kew library that became one of my favourite resorts. I even bought myself a season ticket to the botanic gardens so that I could walk through them on my way to or from a day of study, or drop in for a lunch break to walk beneath the Indian horse chestnuts or hear a blackcap sing in the limes or watch a cherry blossom spring give way to a rose garden summer. Always refer to it in full as the 'Royal Botanic Gardens, Kew', the press office had insisted, and so here, I dutifully do.

Then Covid came, as it came to us all, and for months on end it put a stop to all this. The libraries and the galleries and the archives were closed and my reading and researches had to continue at home, entailing an extra expense in book purchases that my publisher's advance has so far failed to take account of. It was nonetheless increasingly eclectic as I was led to both science and science fiction or ranged from medieval herbals to contemporary astrophysics, from metaphysical poetry to Elizabethan plays. Due to the new-found, and rather mixed, benefits of Zoom, I was able to continue conversations and interviews, including some with experts in fields that I hardly knew existed. It was nonetheless a wonderful release when I was able to resume the expeditions that had always been the happiest part of my researches; the stumbling about, often accompanied by friends or family, in woodland, hill and heath. In the Highlands of Scotland, at the head of its longest glen and in the shadow of the Schiehallion mountain, I visited Europe's oldest tree. On the Hebridean island of Iona, I inspected lichen on the graves of kings, then travelled to the humbler island of Hayling, on England's south coast, where I sought out the last remains of a one-time mosquito research institute. I walked the Hoo peninsula in Kent and the remote coastline of Dengie in Essex, scoured ancient woodland in search of rare lichen, visited the rock-hugging yews of Wakehurst and looked for star jelly sites near Neolithic monuments in Wiltshire. Exciting though these were, they were no more significant than the explorations made immediately beyond my own front door. On these 'excursions for everyone' I searched for surprising plants in my local park, walked surrounding streets surveying mosses, examined the lichen of pavements and walls, and scraped slime from the bark of city trees to take home and sieve for tardigrades.

Here then is my assembled 'casebook'; the story of the missing fragrance of the musk and of the ancient and unsettling associations of the yew; of the unexplained appearances of star jelly and of the secret identity of lichen; of the uncertain origins of an underground mosquito and the possibly astral ones of the tardigrade. I had, I suspect, first set out on my investigations rather in the spirit of a character from an Agatha Christie book. Though my tastes in detective fiction run more to Michael Dibdin, Sara Paretsky and Tony Hillerman, for this work something more traditional had seemed appropriate. I saw myself adopting the manner of a Miss Marple or a Hercule Poirot, following up clues, visiting the scene of a crime and spotting what no one else had. I would amass information and interrogate facts. I would assemble competing theories like suspects in the drawing room, before swinging round to point the accusing finger. I would, in short, crack the case wide open. And then, as these pages will show, something different began to happen. I began to respect the mystery.

The Missing Musk

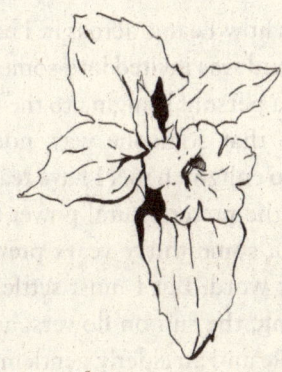

It was 'A World Mystery'; or so ran the headlines in the *Daily News and Chronicle* in September 1930. The event itself had happened some years earlier but, overshadowed by the outbreak of the First World War and the dreadful flu pandemic that followed, it had been largely overlooked. Now it was back in the news. In 1913 all the musk plants in the world had stopped smelling. 'If I were asked to say what is the most mysterious thing that has happened in recent years,' wrote the bemused reporter, 'I think I should answer the simultaneous loss of scent by musk plants all over the world. It was as if by some far-reaching conscious process this charming little flower suddenly decided to deprive mankind of the pleasure of its fragrance.'

It was a fragrance that had once been so intrinsic to the character of the plant that it had given rise to both its English and its scientific names. It was *Mimulus moschatus*, the musk plant. And now, not only had it lost that fragrance, it had done so with amazing rapidity. Once it had resurfaced, the story continued to command attention. It was reported in newspapers, referred to in books, discussed in learned symposia and filled the correspondence columns of leading journals. And when it faded from public consciousness, as even the best stories will, it survived in folk memory and in oral tradition, a sort of rural 'urban myth'.

That, at least, was how I came across it. I had been walking in the Yorkshire Dales and was invited into someone's home for tea and biscuits. It adds a personal enigma to the story that I cannot now remember who that someone was, nor have subsequent enquiries been able to enlighten me. I have read autobiographies whose authors have the preternatural power to recall conversations that took place some thirty years previously, replicating them word for exact word. But I must settle for hazy recollections: a garden setting, the sun on flowers, a china teapot on a circular cast-iron table and an elderly gentleman who told me he had once been something senior, perhaps even Head Gardener, in an urban parks department somewhere. And then this pronouncement: that in 1913, all the musk plants in the world had stopped smelling.

Since he had once been something senior somewhere, and moreover had a beard, I had no reason to doubt his authority. Nonetheless, I checked the story as soon as I got home, taking out my copy of the Reverend W. Keble Martin's *Concise British Flora*, a book that had been my companion on almost every country outing since I was eighteen. Keble Martin was perhaps the last representative of the great English tradition of parson naturalists and his book contains over a hundred plates, packed with his own hand-painted illustrations. There, amongst the accompanying descriptions, was mention of the musk, and the words 'formerly fragrant'.

Laying aside the doubtful usefulness of this phrase in a book that was devoted to identification, I turned to the book next to it on the shelf, Clapham, Tutin and Warburg's *Flora of the British Isles*. This was a weightier and more earnest tome. Where Keble Martin had been a friend amongst botanical books, this one was more of a father figure. Its 1,591 pages did not descend to the frivolity of illustration, unless they were detailed line drawings demonstrating differences in an obscure point of identification. CTW, as it was generally known, was the botanical bible of its day, and not just the bible but its Authorised Version. Here, on page 871, was my plant and, at the end of its scientific description, the words, 'Formerly much cultivated for the musky scent of all parts of the plant. All the plants in this country today, however, appear to be scentless.'

Here then, from my own bookshelves, was validation, at least in part, of what I had heard in that Yorkshire garden. I determined there and then to investigate. Life, however, has a habit of interfering with intention, and, apart from a few early phone calls, it was a full thirty years before I got round to it. But stumbling across the plant during a walk in Glasgow one summer's evening, all the questions I had asked came back to me. What was the story behind the musk and why had it lost its scent so suddenly? Why had the pre-apocalyptic year of 1913 been singled out for this occurrence? What had the plant once smelt like? And from where had it first come? It was with this last question that I decided to begin. It proved to be a detail that was significant in unravelling the rest of the story; and one that would turn up an unexpected mystery of its own.

* * *

In 1824 the plant collector David Douglas began travelling in the remote and rugged regions of north-west America. It was a land of high mountains and large forests, of stony passes, deep ravines and rocky torrents. In the three years he spent there he coped with the constant presence of bears, with hunger and cold, with the occasional hostility of the native peoples, with icy winds and with freezing rivers which he several times had to swim. The domestic connotations of the contemporary garden, the shrub beds and the bedding displays, the neatly mown lawns, the Sunday afternoons with *Gardeners' Question Time*, make it hard to imagine the courage of the early plant collectors, and the privations they put themselves through in order to populate our suburban plots.

Douglas alone was responsible for introducing 254 plants to this country. They included species of lupin, delphinium, honeysuckle, hyacinth, berberis, antirrhinum, evening primrose, potentilla and phlox. The shrub *Garrya elliptica*, with its long, drooping, silver-grey spring catkins, that so resemble the tassels on some ornate cushion or curtain, is one of his plants. So, too, is the flowering currant with its leafy summer fragrance, the now ubiquitous snowberry, with its globular white berries borne long after the leaves have fallen from the bush, and the yellow-stemmed dogwood with its startlingly bright winter twigs. Recently, at the end of a christening, I was, along with all the other guests, handed a packet of seeds. They were another of his introductions: baby blue-eyes, or *Nemophila*, whose beautiful deep-blue flowers fade to a pure white centre. More significant, however, than all these flowers and shrubs were the trees. Among them were six new species of conifer that changed the face of British forestry. They included the Douglas fir that bears his name, as well as the Sitka spruce, now the most widely planted forest tree in the country. Almost overlooked among all this, he collected and brought back with him the seeds of *Mimulus moschatus*.

Douglas was born in 1799 in the small town of Scone in Perthshire. One of six sons of a stone mason, he was registered at the parish school at Kinnoull. He seems rarely to have attended, preferring instead to roam in the surrounding countryside, fishing or hunting for birds' eggs. At the age of eleven he was apprenticed to the Head Gardener at Scone Palace, seat of the Earl of Mansfield, and later went on to work as an undergardener on the Valleyfield estate near Culross. It was, however, his move to the Glasgow Botanic Gardens in 1820 that determined the future course of his life. Here he met William Hooker, then Regis Professor of Botany at the University, and one of the greatest scientific minds of his day. The two formed a friendship that was unlikely, given their difference in social standing, and Douglas became a 'favourite companion' on Hooker's botanical excursions into the Highlands, excursions on which they would sometimes walk as many as thirty miles in a day.

Hooker was later to write, in an appreciation of Douglas, that his 'great activity, undaunted courage, singular abstemiousness and energetic zeal, at once pointed him out as an individual eminently calculated to do himself credit as a scientific traveller'. It was no doubt these virtues that led Hooker to introduce him to friends at the Horticultural Society of London, the forerunner of the modern Royal Horticultural Society, and through this to his first foreign trip. It was intended that he would go to China but this was cancelled when diplomatic relations suffered a breakdown, and he set off instead on a four-month trip to the north-western states of what is now the USA and British Columbia.

Such collecting trips were, in a more or less conscious way, part of the great imperial project, opening up the world to Western science and exploring and exploiting its natural resources for domestic use. But this was probably far from the mind of Douglas when he set out on his second and much longer trip in 1824, returning to north-west America and using a base on the Columbia River to explore the then independent states of Washington and Oregon. By the end of 1825 he had covered 2,105 miles on horseback or on foot. In 1826 he added another 3,932 miles and, by the autumn of October 1827, had completed the incredible feat of walking all the way to Hudson's Bay on the opposite, east coast of the continent. It was from here that he returned to Portsmouth on 15 September, having, as he put it, 'enjoyed a most gratifying trip'.

Douglas kept diligent field notes during the day and, some-how overcoming the exhaustion of each day's exertions, wrote them up in a journal in camp in the evening. Learning that they still existed and were housed in the collections of the Royal Horticultural Society, I determined to see them. So it was that one hot July afternoon I made my way to their headquarters in Vincent Square, one of the largest London squares set between fashionable Pimlico and the more downmarket bustle of Victoria Station. It is a building of suitable gravitas for a royal institu-tion, though the severity of its red-brick frontage is leavened by parallel bands of white stone, rows of stone-framed windows and a rather jaunty gabled roof. Entering through its imposing portico I was directed to the Lindley Library, occupying one of the ground-floor wings. Here, beneath the severe countenances of past RHS presidents, I was introduced to the archivist. She was not severe at all and led me downstairs to a large room where controlled temperatures kept conditions at a level just below comfortable – 'bring a pullover', I had been advised when ringing to make the appointment.

The journals, in four volumes, were brought out on a trolley and lifted with reverence onto cushions on the table. I had hopes of being given a pair of delicate white gloves with which to examine them but, in this, I was disappointed. The volumes were in varying condition, but all were beautifully bound in boards marked with a shell pattern, blue-veined on a base of red and brown. The script was in a fine sloping hand in a black ink that was, in most places, fading to brown. Having opened a page at random, the very first line I encountered read, 'No sleep until completely worn out with fatigue'. It was not always easy to decipher the script and in some places there were signs of haste with frequent crossings out and underlinings. The reverse of some pages were written on, upside down, as though he had come back later to find extra space, while some volumes had additional pages pasted in at the back.

Hoping to pinpoint the moment when Douglas had collected the musk, I ploughed through them with the help of a later typed transcript. On 4 August 1826 he had been thrown from his horse while crossing a river and immediately after that, in an entry spanning the next two days, I found a possible candidate. 'In addition to a good many seeds,' he had written, 'collected the following plants on my journey: *Mimulus sp.*; annual, leaves alternate, nearly sessile, slightly dentate, stem pubescent, flowers small, yellow: on moist rocks and mountainous grounds'. It is a description that fits closely to *Mimulus moschatus* except for the reference to the plant being 'annual'. *Moschatus*, like all the other species of *Mimulus*, is a perennial, but it is a short-lived one; a low scrabbling plant that soon dies back leaving behind just its dried seedheads. The misattribution, I thought, was understandable. Even though he had collected a total of five different species of *Mimulus* on this one trip I was fairly confident that 5 August 1826 was the day he had first come across, and collected, the seeds of the musk. What he had not mentioned anywhere in his journals, however, was that it had a fragrance. It was an omission that was later to prove significant.

The journals were not the only documents I was handed that afternoon. Carefully wrapped in crepe paper, another set of notes was brought out from the archives. They were handwritten on roughly cut sheets, bound into loose bundles and fixed with ribbons and pins. In the same neat, leaning hand, but this time with fewer corrections, these were more detailed notes on the plants he had collected, written up by Douglas in the two years following his return. There, among them, was what had now been named *Mimulus moschatus* with its Latin description: '*Caula repente, foliisque glandulosa-villosis . . .*' And after this, in English, a note on where it had been found: 'On the margins of grassy springs', ending with the words, 'when trod on emits a powerful odour of musk'. The words seemed to establish what the journals had failed to: that the musk introduced to Europe by David Douglas was indeed fragrant. The story of the loss of that fragrance was still to come. The story of David Douglas, meanwhile, was moving on to its own mysterious conclusion.

* * *

A transcript of the Douglas journals, produced by the Horticultural Society a few years after his return, contains as its frontispiece one of the very few pictures we have of him. It is a pencil drawing 'by his niece Miss Atkinson' and shows a clean-shaven and rather good-looking young man in a high-collared coat, his lips full and slightly curled, his hair curly, his expression pleasant. But Douglas was not always an easy man. He could be impatient and ill-tempered, often falling out with even his friends and patrons. On his return to Britain, bearing his journals, his seeds, his specimens and his stories, he had been met with a hero's welcome. For a time he was lionised, but he soon tired of these attentions. As Hooker put it, with his usual forthrightness, his irritability was such that 'his best friends could not but wish, as he did, that he were again occupied in the honourable task of exploring North-west America'. Which is what he very soon set out to do.

In 1831 he began his third and final trip, travelling back to the north-west states where he spent another three years collecting, surviving another series of adventures which included capsizing his canoe and nearly drowning whilst navigating rapids on the Fraser River. He began his journey home via Hawaii, then known as the Sandwich Islands, which he had also visited on his outward trip. Among his many accomplishments Douglas was one of the earliest mountaineers and within a single month he had climbed the island's three highest peaks, reaching the summit of the 13,679 ft Mount Loa, on 2 January 1834. In May of the same year he wrote his final letter to Hooker: 'May God grant me a safe return to England I cannot but indulge the pleasing hope of being able, in person, to thank you for the signal kindness you have shown me . . .' But he was to see neither Hooker nor England again.

In July, following a delay in the arrival of his boat, he began traversing Hawaii on foot. He was accompanied by 'John', a servant lent to him by one of the missionary families on the island, and by Billy, his small Scottish terrier. Setting out early on the morning of the 12th, he stopped at the hut of Edward Gurney for breakfast. Gurney, generally known as 'Ned', was an ex-convict, an escapee some said, from the penal colony at Botany Bay, who now made a living trapping wild cattle. The technique was to dig pits on the bullock trails and cover them with foliage and branches. The cattle that fell into them would then be shot. According to Gurney's testimony, he urged Douglas to await the arrival of local guides but when he proved impatient to get going, Gurney himself accompanied him for a mile or so along the trail, pointing out the location of the bullock pits along the way.

What happened next is uncertain. What is known is that local people following the trail later in the day noticed a torn piece of cloth above one of the pits and, looking into it, saw a trapped bullock standing on the body of Douglas, who had been gored to death. They called for Gurney who shot the animal and retrieved the body. From there it was transported by canoe to the missionary station at Hilo, where Douglas had been heading. The fullest account of the incident is contained in a long letter, written to the British consul at Honolulu, by John Dill and Joseph Goodrich, two of the missionaries at the station. It is couched in the florid tones characteristic of the time:

Dear Sir, Our hearts almost fail us when we undertake to perform the melancholy duty which devolves upon us, to

communicate the painful intelligence of the death of our friend
Mr. Douglas . . . The tidings reached us when we were every
moment awaiting his arrival, and expecting to greet him with a
cordial welcome: but alas!

After a paragraph of pious reflections about the immutability of
God's purposes, illuminated by appropriate biblical texts, the
letter goes on to describe the arrival, and then the examination,
of the body: '. . . what an affecting spectacle was presented as we
removed the bullock's hide on which he had been conveyed! – we
will not attempt to describe the agony of feeling which we experienced at that moment: can it be he? can it be he? we each
exclaimed.' Their initial thought was to give him a decent burial
but it was then that the first doubts about the exact nature of his
death began to surface. The wounds on Douglas were numerous, but at least two of the people who inspected the body
doubted that they could have been inflicted by a bullock. After
enumerating some other concerns they decided to send a couple
of 'foreigners', by which they meant Americans, to investigate
the site.

The letter resumes on the following day, in the course of
which Edward Gurney himself arrives at the station. He seems
to have exerted some charm over the missionaries for the letter,
having previously named him as 'Edward Gurney', now switches
to the more familiar 'Ned'. 'Our minds', they say as he provides
his own account, 'are greatly relieved'. They determine, nonetheless, that the body should be examined by 'medical men' and
prepare it for transit to the consul. The contents of the abdomen
are removed and the cavity filled with salt. It is placed in a coffin
stuffed with more salt and the whole encased in a box of brine
which, together with the missionaries' written account, is sent
on its way by boat to Honolulu.

Douglas was as well known to the consul as he was to the
missionaries and he is deeply distressed at the sight of the body.

As he reports in one of his own letters, 'I assure you that I scarcely received such a shock in my life ... On opening the coffin, the features of our poor friend were easily traced, but mangled in a shocking manner, and in a most offensive state.' He immediately has the body examined by the doctors who give it as their opinion 'that the several wounds were afflicted by the bullock.' And with that assurance formal investigations into the case seem to have come to an end.

But doubts remained, and over time they multiplied. How could an adult bullock fall into a pit and not leave a hole in the covering large enough to be seen by a passer-by? Why had it taken Gurney four days to turn up with his account of the death of such a well-known person? And what had happened to the servant John? Accompanying Douglas up to this point on his journey, he receives no further mention in any of the accounts. He simply disappears.

Then there is the matter of the bag that Douglas had been carrying. It was found not in, or by, the pit, but some distance further up the trail, the suggested explanation being that he had for some reason put it down and begun to retrace his steps. The contents of the bag raised a further question. Douglas had lodged the previous night with a Mr Davis who stated that while being paid he had seen the amount of money that Douglas was carrying. It was, he claimed, far more than was subsequently found in the bag which Gurney had passed on to the missionaries.

If Douglas was indeed murdered, as some were now openly suggesting, then robbery provided a motive. In subsequent years another was to emerge. A letter to the Honolulu *Star Bulletin* in 1949 cites a view that Douglas had been in a relationship with Gurney's wife. That this was common gossip amongst the native peoples was demonstrated by papers found, much later, in the collections of the Edinburgh Botanic Gardens. This 'relationship' may have amounted to no more than the fact that Gurney's wife preferred the charismatic and attractive foreigner to her own husband, but it adds jealousy to the story's already complex mix.

Douglas was much loved on the island and the missionary families, in particular, looked forward to his visits. Their letters and journals make it clear that they found it very difficult to believe that such an experienced mountaineer and explorer would have made the mistake of falling into a bullock pit. The native peoples, many of whom were afraid of Gurney, shared this view. A letter to Hooker by a Dr Meredith Gairdner, an English visitor to Hawaii, suggests that even the consul may have had his doubts: 'the minds of the residents and particularly of the British Consul . . . [are] by no means satisfied that he came by his end by mere accident'. By 1856 a Californian newspaper was openly reporting the possibility of murder, and citing the suspect. In words translated from the French, it read:

> We have received a white marble monument from San Francisco, erected by Mr. Julius L. Brenchley to the memory of the illustrious and unfortunate voyager David Douglas, who died in 1834 at the foot of Maunakea, on the island of Hawaii. According to some he was murdered by a convict escaped from Botany Bay; according to others, killed by an enraged wild bull.

But what of the 'foreigners' dispatched by the missionaries at Hilo to investigate the scene? Strangely, I could find no contemporaneous account of their findings: no written report, no transcribed conversations, no hints in journals or letters. In fact we have to wait twelve years until a Chester Lyman, visiting Kona, on the west coast of Hawaii, meets with the coffee planter Charles Hall. Hall, it transpires, had once been a bullock hunter himself and was one of the two Americans dispatched to undertake the investigation. Lyman records something of their conversation in his journal:

There had been a heavy rain the day before he reached the place and all tracks were obliterated. He obtained the head of the bullock and took it to Hilo. The horns he says were blunt and nearly an inch through at the extremities, the animal being old. He thinks it impossible that the wounds on Douglas's head especially on the temple . . . could have been made by the horns or hoofs.

It is fair to say that the accident theory still has its supporters. John Davies, who produced an edition of the Douglas journals in 1980, suggested that having passed the pit Douglas then heard the sound of a bullock trapped within it. As insatiably curious as ever, he put down his bag and walked back to take a look. Already short-sighted, he went too close to the edge and slipped as the ground gave way, falling to an accidental, but horrific, death. It is a reasonable theory but I remain unconvinced. The evidence is entirely circumstantial, and indeed, at this distance in time it can hardly be otherwise, but there are too many doubts, too many discrepancies and too many niggling details. Douglas, I believe, was murdered, and for that murder, Edward 'Ned' Gurney has a significant case to answer. But, in truth, we will never now be certain. As a report in the 1988 edition of the *Journal of Hawaiian History* concluded, 'The many accounts of Douglas's death that circulated over the years do not clarify what actually happened to him. They merely deepen the mystery.'

Douglas was only thirty-five at the time of his death. He was buried in an unmarked grave beside Kawaiaha'o Church on the island. Later the Royal Horticultural Society erected a plaque there, and a cairn was built close to the site of the pit in which he died. Yet his real memorial remains the many plants that bear the scientific name 'douglasii', the dozens of his introductions that grace our gardens, and the thousands of Douglas fir and Sitka spruce that fill our forest plantations. In the course of writing this account I discovered that there is, in fact, a Douglas fir-flavoured gin. It seemed only appropriate to order a bottle, invite a few friends round and toast him.

* * *

When the samples from the 1824 expedition arrived at the Horticultural Society, the work began of naming and describing them. In accordance with taxonomic principles, the full scientific designation applied to the musk was '*Mimulus moschatus Douglas ex Lindley*'. The appendages denote that though the plant is attributed to Douglas, the authority for it, and for its first detailed scientific description, belongs to another man: John Lindley. And with Lindley we come to another significant character in the story. For some time a Professor of Botany at the University of London, he was a man of huge energy, said to work daily from dawn till after dark. He published over 200 books and articles, amassed a herbarium of more than 58,000 sheets and delivered as many as nineteen lectures in a week. 'Nature', he once wrote, 'bears the living hieroglyphics of the Almighty', and perhaps that is what he set out to prove in his work, bringing about major revisions in taxonomy, producing specialist studies of orchids and organising the first public flower show in Britain. Such was his reputation that in 1845 he was appointed by Prime Minister Robert Peel to a scientific commission investigating the causes of the Irish Potato Famine. Though unable to come up with a cure for the blight, his report was to lead to the repeal of the protectionist Corn Laws that had done so much to exacerbate the problem. As a leading light in the Horticultural Society, he bequeathed it his large collection of books. They came to form the basis of the Society's library, the one in which I was then working, and which still bears his name.

Where Douglas could be irritable, Lindley was once described as being 'forthright to the point of appearing brusque'. Perhaps it was inevitable that the two men would fall out, in one of those rows that made Hooker so eager to see Douglas off on his travels again. The cause of their disagreement is unknown but it is very likely that it concerned the cost of the Douglas expeditions, for Lindley had been critical of the Society for spending more than it could afford on what he described as 'grandiose projects'. Perhaps there was some merit to his case, for the Society's Honorary Secretary was later dismissed for the mishandling of the funds, but this would hardly have endeared him to Douglas.

It was Lindley, nonetheless, who took charge, not only of naming the specimens that Douglas had brought back with him, but of growing on the seeds that he had collected. Of the 210 different species sown at the Society's new gardens in Chiswick, eighty were recorded as being merely 'botanical curiosities' and thereafter abandoned. The remaining 130 were considered to have some horticultural value and were distributed to other gardens and to horticulturalists, both in this country and abroad. Lindley had, rather dismissively, described the herbarium specimens of the musk plant brought back by Douglas as 'not very helpful'. It was therefore the live plants grown at the Chiswick gardens which he used to produce the description, no doubt correcting along the way the impression given by Douglas that the plant was an annual. As a result of this work, the disputative pair now have both their names inextricably linked in the nomenclature of the musk.

* * *

From these first Horticultural Society sowings the musk plant enjoyed a remarkably rapid spread. Within a very short period of time it was being offered for sale by nurseries and eagerly sought after by gardeners. The earliest listing I found in the Lindley Library was a seed catalogue produced by Flanagan and Nutting of Mansion House, London in 1835, just eight years after the plant's arrival in the country, though there may well be earlier examples. It was, almost from the first, a democratic sort of a plant and not just one for the better-off suburban gardener.

It was popular for small plots and cottage gardens and among those with no gardens at all, grown in containers, pots and window boxes. It was sold in masses from market stalls, one Liverpool trader in the high season selling 5,000 plants a week. And in London it was hawked from street to street on barrows. By 1844, just twenty years after its introduction, a popular gardening book was giving the following account:

> *A very few years ago the favourite little Musk plant, now so common as a parlour or window flower, was unknown in cultivation; now it grows everywhere as it deserves to do. In the morning its scent is weak and faint; in the evening most powerful yet delicate – a single plant scenting a whole apartment, sometimes even too powerfully.*

Soon it was also part of the colonial enterprise, a kind of nostalgic memento of the mother country that early settlers took with them or ordered from the homeland. And so the musk was carried to New Zealand, to Canada, and even back to the parts of America from whence it had first arrived. According to one account of the 1860s Otago gold rush on New Zealand's south island, where new and bustling towns were springing up almost overnight even before roads had been built to serve them, some of these tough prospectors and their families were importing collections of garden plants, including the musk, by pack horse.

In Scotland, meanwhile, the relationship between the plant and the poorer sections of society was taking another turn. Frances Jane Hope was a pioneering gardener and a woman ahead of her time. Already, in the mid-nineteenth century, she was advocating the sorts of principles, naturalistic planting and complementary colour schemes that would later be made famous by Gertrude Jekyll. She expounded these ideas in more than fifty articles for garden periodicals but also, and almost uniquely, saw her Edinburgh garden as a way of helping those less fortunate than herself. From her grounds at Wardie Lodge she began carrying sprigs of spearmint and rosemary to old women in the cellars and garrets of Cowgate, and collecting posies of flowers and herbs to distribute to the city's sick and poor. She went on from this to develop her 'Flower Missions', eventually delivering up to several hundred baskets a day to the Royal Infirmary and to the Asylum for the Blind. It was here, among the blind women inmates who sorted the flowers and made them into posies for sale, that the fragrant musk plant was particularly popular. Eventually a large part of her garden was totally devoted to the production of plants for philanthropic use, and she became exhausted by her efforts, dying at the age of 58. As Jane Brown put it in *The Pursuit of Paradise*: 'We hardly have to ask why Wardie Lodge baskets carried happy memories and touches of human dignity back into those dreary places but we may be sad that one garden, however abundant, could not supply such overwhelming needs, and that Miss Hope wore herself out, perhaps to her early death, in trying.'

* * *

Clearly the little plant with the powerful scent had meant much to many people. Had the loss of that scent, however, happened quite as suddenly as the popular story suggested? To research that question further I took myself first to the London Natural History Museum where, for several days, I threaded my way through crowds of buggy-pushing parents, excitable school parties and tourists who seemed to be making a photographic survey of the museum's entire collection. The rather secretive library doorway was set between display cases of aardvarks, pangolins and a snake wound round a branch like a rugged Aesculapian staff. In the galleried hall beyond, a statue of Darwin seemed to have obtained a library card before me and to be sitting at study at one of the desks. It was clear he was making better progress than me for I found little of help and my question to the museum's botanists, which I was required to submit in writing, got me little further advice than that I should consult Wikipedia. I turned my attentions instead to the archives at the Royal Botanic Gardens in Kew. The hallowed hall at the Natural History Museum had been window-less, and with that slightly stuffy atmosphere that makes it difficult not to doze off at some point in the day, but this was bright and airy and, better still, situated adjacent to the botanic gardens. I treated myself to a season ticket and combined my working hours with a twice-daily walk. At this time in spring the gardens were achingly beautiful. Blackcaps sung in the still bare trees, and distant green woodpeckers uttered their plaintive yaffle. A huge spreading magnolia bore its fat buds like a giant candelabra, and a whole lawn was taken over by glory-of-the-snow, a mass of pale blue flowers punctuated by the white of daisies. The air was full of fragrance; soft and mellow scents with sharper notes here and there and, what seemed to me, an inexplicable hint of liquorice. But most of all, it was cherry blossom time, that brief enigmatic period so symbolic of transience, when the trees in bright avenues become thick with flowers in white or pale pink; exuberant with them until they are suddenly shed in wind-whipped snow-like showers.

The library staff could not have been more helpful, locating books, digging out files and identifying old volumes of time-battered and brittle-paged journals. It was through the correspondence columns, the 'Notes from Fellows', the learned articles in publications such as the *Gardeners' Chronicle*, *Nature* and the *Journal of the Royal Horticultural Society* that the story began to unfold, helped along by an increasing number of interjections from the national press.

In poring through these, the earliest account of the plant's loss of fragrance I came across was dated from 1898. Though it was not published until the 1930s, it was then that Thomas Wilkinson, the horticulturalist who had been selling the plant so successfully in Liverpool, claimed to have noticed that the musk plants he was stocking had begun to acquire a rank, leafy smell. With an admirable prescience, he sold his entire business as soon as the season was over. Returning to the city some four or five years later, he comments that the plants on other people's stalls were also losing their smell. Another early reference comes from a letter to *The Times*, although again not appearing till the 1930s. Here an appropriately named Mrs Gardner states that when she was living in Surrey in 1903 the bed of musk which she grew every year beneath her front window had come up scentless. The same thing, she claimed, had happened to all her neighbours and friends. It was becoming apparent that while the plant's loss of fragrance was real enough, it had not happened quite as abruptly as the story suggests. The reality was, as ever, more complicated.

Through the first two decades of the twentieth century evidence of the phenomenon began to multiply. Writing to the *Gardeners' Chronicle* in April 1909, Mr T. Smith, another well-known nurseryman, asks, 'Is there such a thing now as a Common Musk with the old Musk perfume? Many friends of mine contend that there is not and I myself am sceptical.' A herbarium specimen of the plant, collected by the Reverend E.S. Marshall in Somerset in 1916, has the word 'Scented' written next to it, and then, in brackets, 'it usually seems to have lost its odour nowadays'. By 1920, Mr Alfred O. Walker of Maidstone is almost pleading with the public in the pages of the *Gardeners' Chronicle*: 'Can any of your readers inform me whether the Common Musk is to be had with the old characteristic smell? I have had the plant (but not the smell) in my garden for many years. I have tried raising it from bought seeds but the seedlings were quite scentless.'

There is a significant point in these accounts when the comments are less on the widespread absence of the scent as on the remarkable discovery of a plant that still produces it. In

October 1930, the *Daily Mail* reports that in Hampshire 'a fully scented specimen' has been raised from seed. The next month, a Mr Chittenden writes more sarcastically to the *Gardeners' Chronicle*: 'Several times since the war we have met someone whose uncle's greatest friend, recently dead, had told him that his butler's niece by marriage had a cousin who had heard of a plant smelling distinctly of Musk growing in a cottage window'. An Inverness gardener states, in 1931, that he can still detect a slight perfume in his plants 'on certain days'. A correspondent known only as 'C.E.' claims, as late as September 1947, to have two pots of fragrant musk: 'It certainly does not scent out the garden, but when I put my nose near it I can get the sweet smell. I do not say it is constant, but neither are a lot more scented plants'. Finally, also in 1947, and before the reports cease altogether and public attention turns elsewhere, Mr Henry Ridley reports in the RHS *Journal* that he has been sent from Kew to Parkstone in Dorset to meet a Miss Claydon. Her distinction is that she has grown pots of musk with 'a distinct, fairly strong scent . . . reminding me of plants of the 1860s and 1870s when every cottage garden had clumps of this plant'. But the plants are only one and a half inches tall and by the time they have reached full height, their scent has completely faded. Mr Ridley takes some of them back to Kew with him but, as he sadly reports, 'the scent was gone, never to reappear'.

* * *

In all of these accounts what remains conspicuous by its absence is any actual reference to the year 1913. Yet that remains the date of the popular story and continues to be repeated even in contemporary texts. Thus the gardener and writer, Roy Genders, states in his *Scented Wild Flowers of Britain*, that the event 'happened throughout the world at the same time – in 1913'. And the entry in Wikipedia, consulted as the Natural History Museum botanist suggested, gives the same account. The musk, it says, 'is well known for the story that all cultivated and known wild specimens simultaneously lost their previous strong musk scent around 1913'. Occasionally a source will slip a year, the *RHS Dictionary of Gardening*, for example, setting the date at 1914, as do the authors of *Plants from the Past*, exclaiming that, 'Inexplicably, around 1914, the scent disappeared from every plant in the world'.

Why then this common and incorrect attribution of the loss of scent to a specific year? The artificial contraction of the event and its depiction as sudden, simultaneous and global, has turned a genuine mystery into an urban myth. All the classic ingredients are there: the element of mystery with perhaps a hint of the paranormal, the spread of the story from mouth to mouth and its take-up by the popular media. Placing it immediately before the outbreak of one of the world's greatest conflagrations adds another element, and transforms it into something of a morality tale. The musk plant lost its smell not suddenly but over a period of fifteen to twenty years preceding the First World War. The words 'before the war' feature in several of the accounts and are crucial, I suspect, to the way the story developed. The 'Great War' as it was known at the time, was one of the dividing points of history, a descent into an unimaginable and self-inflicted horror. Nothing after it would be the same. An era, a particular order of society, even a way of seeing the world, had come crashing to an end. It was a second expulsion from Eden and everything would now be dated as occurring before or after it. As the story of the musk was passed from person to person it was inevitably dramatised; that, after all, is the essence of storytelling. The twenty- or thirty-year period in which the scent was lost, itself a remarkably short period of time, became compressed into a single year. And the fact that it had happened 'before the war' became literally that; the year that preceded the outbreak of hostilities, or, in some cases, the very year that they had started. And this served too, to give an already dramatic story a chillingly apocalyptic twist. Just as comets had once been harbingers of terrible events to come, so this universal loss of scent is seen somehow to presage the disaster. As if in a Greek myth, the little flower has renounced its fragrance in grief or despair, or in terrible prophecy of the coming decimation of 'the flower of Europe's youth'.

* * *

Gradual or precipitate, its loss of its fragrance, was to seal the fate of the musk. It had none of the other features that lead to popularity in a garden plant, and its demise was as rapid as had been its rise. It was, as the reporter on the *Daily News* commented, as though it had 'passed sentence of death on itself, for no one wants it without its scent, and it is now only a fragrant memory to most of us.' It was a horticultural extinction, self-inflicted or not, that is confirmed by any modern gardening book. The 2016 edition of the *RHS A–Z Encyclopedia of Garden Plants* runs to 1,120 pages. It contains no mention of the musk plant. Even longer at 1,583 pages, the 2003 edition of *Flora: The Gardener's Bible* is similarly silent. *The Flower Expert*, part of the hugely successful gardening series by Dr Hessayon, lists eight species of mimulus as suitable for cultivation. *Moschatus* is not among them. The musk, as a garden plant, has simply disappeared.

It was with a rather perverse determination, therefore, that I set out to see if I could grow it. Obtaining seeds would be the obvious problem. There were none in the catalogues and when I did finally locate some, they were on eBay and had to be sent from China. When the eagerly awaited package arrived it was not surprising that the instructions were in Chinese, but there was a small panel with an English translation of sorts, and one which seemed to be hedging all its horticultural bets: 'Perennial herb. Annual biannual herb. 25–35 cm height, caudex sturdy, Funnelform, suitable to garden, resistant cool weather but can not be freezing injury. Like moist, sunny.' Though the Latin name was clearly given as *Mimulus moschatus*, the picture on the front showed a flourish of radiant, multi-coloured blooms, in yellow, white, pink and a deep velvety red. The biggest surprise came when I opened the packet. Inside was a tiny, sealed, cellophane envelope, and it seemed to be entirely empty. Knowing that mimulus seeds are very small, and that my eyesight is not now of the best, I hopefully wafted the empty bag over a tray of carefully prepared seed compost. It was to no avail. If there were invisible seeds within, then it was to invisible plants that they gave rise.

Anyone wanting to see the musk in Britain today must turn from the garden to look in the wider countryside. During the period of its popularity, seeds spread from cultivated plants to naturalise themselves in wild and semi-wild locations. It is not, however, an easy matter to find them. Described as growing in moist, grassy places, preferably with a little shade, the musk plant occurs on the muddy edges of ditches, in damp woodland rides, beside ponds and in wet pastures. But, in keeping with such an enigmatic plant, it is scattered and elusive and frustratingly irregular in its occurrence. It is a plant you seem to have to stumble across, rather than go out and look for. My first encounter with it was in the scruffy corner of a Suffolk caravan site, but the sites I most associate it with are in Glasgow. Every year I visit the city and stay with friends in the West End. And every year I take a stroll along the River Kelvin. It is a fast brown river, running past tall blocks of flats on the city's outskirts then cutting its way through urban Glasgow and the Kelvingrove parks until it reaches the Clyde. Its deep, damp valley is thick with dripping trees, with the lush growth of ferns and with the plants that typically haunt a post-industrial landscape – giant hogweed, Himalayan balsam and greater knotweed. The path runs under finely engineered bridges, alongside abandoned mill leats and past a weir where a hunched and untidy-looking heron keeps regular watch. There are usually goosander about too, a bird that tolerates the torrents, beautifully slim and streamlined and with a long, thin, fish-seizing bill. For some of this stretch the river runs directly beside the Glasgow Botanic Gardens. It is not the same site as the one in which Douglas worked in the 1820s, for it moved to this place some fifty years later, but it is the same institution, and it seems appropriate that it is near here that I have several times found his plant. It was there in puddled places beside the path, or huddled at the base of a line of railings; a rather scrabbling little plant which seemed to have been humbled by the loss of its famous fragrance and now knew its proper station. Its jumble of hairy stems bear unstalked leaves, lightly toothed and a pale, and sometimes almost anaemic, green. From this tangle, the flowers emerge on short stems, a cheerful buttery yellow with red veins in their furry-looking throats. The five petals form a particularly distinctive shape; two at the top with two, slightly larger, at the sides, and an odd one at the bottom, projecting a little, like a lip, or perhaps a chin. They remind me of the gaping mouths of fledglings demanding food when the parent bird returns to the nest. To some, perhaps more classically trained, they resembled the masks worn in ancient Greek drama and from this came their scientific name of *Mimulus*, derived from *mimus*, the Latin for an actor. The English name for the genus as a whole is monkey flowers. It is said to come from the resemblance of the flower to the face of a grinning monkey. A more likely explanation, I suspect, is to be found in the traditional idea of the monkey as *mimus*, the chief of mimics.

* * *

What had it been like, this lost but once vaunted fragrance that had sold the plant in its tens of thousands, that had seen it bound into posies, that was strong enough to scent a whole room? And was there any way that I could still experience it today? Douglas had borrowed his name for the plant from a substance well-known in perfumery. It was perhaps much better known in his day than ours, for it had been one of the most expensive commodities in the world, more valuable by weight than gold. Its origin was unglamorous, however, being extracted from the genital glands of one of the seven species of musk deer that inhabit the mountainous and forested regions of central Asia. Small and secretive, they have the peculiarity of lacking antlers, the male growing instead a pair of projecting tusks, leading one commentator to describe them as 'having the face of a kangaroo and the fangs of a vampire'.

The musk is located in a small skin gland close to the male genitals and is produced in the rutting season to attract the females, a device that is perhaps essential for a skulking creature in deep undergrowth leading an otherwise isolated life. And it is for this unlikely product that the animal was hunted, sometimes almost to extinction. The glands, known as pods, were stripped and removed, then trimmed and dried. Musk removed from the pod was known as 'grain musk' and came in a number of varieties of which 'Tonquin' was the most highly prized. Still highly priced on the black market, it was once at the centre of a massive and lucrative trade. Since it took the deaths of 160 deer to produce a kilogram of musk, the slaughter was massive. Writing in 1913 in his book *A Naturalist in Western China*, E.H. Wilson states that as many as 60,000 of these pods would be exported annually from that region alone. Yet the smell it produces is far from delicate and is said to have an overbearing animal tone. When well diluted, however, this becomes subdued, and reveals an underlying floral note. Added to compositions, including those based on synthetic compounds, it has the effect of rounding off the odour and endowing it with naturalness and warmth. Just as importantly, musk has the quality of restricting the rate at which more volatile compounds evaporate. It is, in other words, a fixative, making the perfume it is a part of last, and linger, longer. But there was another quality, adding significantly to its value, and that was its reputation as an aphrodisiac.

Even the name of the musk has a sexual connotation. It can be traced back as far as ancient Sanskrit and the word *muška*, meaning testicle, from a resemblance of the glands in which it was found to that part of the male anatomy. It came into European languages via the Latin *mus*, which also means mouse since, in someone's imagination, the testicles resembled a couple of nestling mice. It has, in disguise, infiltrated the English language in a dozen different ways. It is there in the muscat grape, the Muscovy duck, the mouse and its 'musty' smell, and even, much disfigured, in the name of the nutmeg. Perhaps not surprisingly, given its origin, the smell of the musk is said to closely resemble testosterone, and to act as a pheromone in humans. With these associations, with its floral and honeyed tones mixed with that essential animal hint, the musk has not only been the fragrance of ancient royalty but also of the harem and the boudoir.

In 1887 a man called Arthur J. Stace came up with an intriguing idea for creating a 'systematic nomenclature for the odours of plants'. His proposal was for a system in which plants could be classified by the effect of their smell on different parts of the body. It would, he argued, also help with their identification. The obvious example was the 'alliaceous' plants, the members of the onion family, which stimulated the 'lachrymal glands', but this principle could, he argued, be extended to other families. The parsley, he suggested, along with other 'umbellifers', principally affected the salivary glands; the 'crucifers', the family which includes the mustards and cresses, 'excited thirst'; the mints 'cooled the mucous membranes'; the absinthe and the wormwoods 'provoked perspiration' and the smell of night-flowering plants led to enervation and headaches. Musk was in a class of its own. 'The odour of musk', he argued, 'represented in the vegetable world by *Mimulus moschatus*, is a notorious aphrodisiac . . . Scarcely anybody will acknowledge that they like the smell . . . nonetheless the perfumers regard it as a principal source of profit'. Here then, albeit in a theory that was almost instantly forgotten, the fragrance of the musk plant had sexual excitation as its defining and identifying characteristic.

This was clearly a scent to search for, and if I couldn't find it in the *Mimulus* I thought I might try among the other plants that shared the word 'musk' in their name. The musk rose seemed a good place to start. Although not perhaps as ancient as the perfume, it has a respectable pedigree of its own, having been around since at least the sixteenth century. And more significantly, it shares the perfume's sensual associations. It grew, in *A Midsummer Night's Dream*, on the thyme-blown bank where Titania, Queen of the Fairies, slept, lulled by 'dances and delights' and by the scents of flowers. Here, in Shakespeare's most magical, mischievous and sensual play, she courted the donkey-headed Bottom and was pursued by the haughty Oberon in a bower:

> *Quite over-canopied with luscious woodbine*
> *With sweet musk-roses and with eglantine.*

It was beneath the musk rose, too, that Lorenzo met secretly with the beautiful Isabella, in Keats' poem of that name. His fate was to be murdered by her brothers, and to have his head hidden in a pot of basil, but for now, having declared his love,

> *All close they met, all eyes, before the dusk*
> *Had taken from the stars its pleasant veil*
> *Close in a bower of hyacinth and musk*
> *Unknown of any, free of whispering tale*

My regular walks through Kew gardens made the task of locating it easier for there is, in the area behind its glorious Palm

House, what must be one of the largest collections of roses in the country. Radiating from a central circle, they are laid out in perhaps as many as fifty beds, with eight to ten varieties in each. And every one of them is labelled. Walking back from the library one afternoon I began my search by circumnavigating each bed and reading the labels. After the first two or three I was hardly seeing the shrubs at all, lost as I was in the names of floribundas, polyanthas, grandifloras, hybrid teas, shrub roses, rambling roses, miniatures, multifloras, English roses, China roses, bourbons, damasks, noisettes ... I began to feel dizzy, and just slightly demented. It was a foolish persistence to keep going, I thought, after twenty or thirty beds, and one which well displayed my obsessive tendencies. Though it would have seemed a clumsy artifice in any self-respecting novel, I was to find the musk rose in the very last plant in the very last bed that I came to. It was there in the south-west corner, in bed number 171–12, close by the statue of a rampant, heraldic, white greyhound, which seemed to be rearing up in either triumph or derision.

Over the next few months I regularly revisited my *Rosa moschata* to watch it grow. From its pruned winter shape it developed into a sturdy shrub with stout stems and few but large and fearsome thorns, coming into leaf behind the other roses but with a fine pale, pea-green foliage when it did. And then, in June, I was back again, to see and to smell it in flower. A light rain was starting as I walked to the beds behind the Palm House and, even though the blooms were not yet at their best, what greeted me was stunning. If I were to list the must-see sights of London, this would surely be among them. Bush after bush was literally bowed down with blossom, weighted with it, in red, white, pink, purple, orange and a genteel faded mauve. There was, it seemed to me, a distinct air of decadence about the place; of decadence and of decay. Living bright blooms were mixed with those already shedding their petals, with those gone over, with those stained brown, with those that were rotting on their stems in the rain. Petals, even as I watched, fell to the ground in great wasteful handfuls. I was immediately reminded of Hardy's wonderful poem 'During Wind and Rain' where the rose becomes a symbol of death and decay:

> *Ah, no; the years O!*
> *And the rotten rose is ript from the wall.*

Others have used it with the same intention, as in Tennyson's poem 'Maud', where the rose, and its scent, appear as a symbol of sexual longing but one that elides with a longing for death. 'Come into the garden, Maud,' he implores her, where,

> *. . . the woodbine spices are wafted abroad,*
> *And the musk of the rose is blown.*
> *For a breeze of morning moves*
> *And the planet of love is on high*
> *Beginning to faint in the light that she loves*
> *In a bed of daffodil sky.*

And, indeed, the musk of the rose was blown in Kew that day. The mingled scent of massed blossom drenched the air so that you could not but breathe it in as you walked. It spoke of something refined but with a hint of rottenness beneath its surface. It was Persian ice cream and rose water and petal sandwiches. It was crinoline and damask and patterned screens and bedrooms with their heavy curtains drawn. It was the bedroom of a frail and ancient maiden aunt with an old-fashioned scent bottle on a dainty dressing table, all 'chintzy, chintzy cheeriness' and with death hovering not far outside the door.

The musk rose itself, however, was different. It was, to begin with, a more unassuming plant than its shrill and showy neighbours. Its single, five-petalled white flowers looked rather vulnerable, opening almost flat with a mass of projecting yellow stamens. It was these that produced the fragrance, which I had to lean in to detect. It was quite different from those that perfumed the air all around me. It was tangy and spicy, with a distinct, and rather surprising, hint of cinnamon, and perhaps, of nutmeg too. It was punch and mulled wine and fruit cup and I loved it. Whether it could be described as musky or not, was another matter.

Not wanting to trust in my opinion alone, I decided to undertake a totally unscientific experiment. I would stop the next ten passers-by and ask them to smell the rose and give me their own description. It would be a sort of olfactory version of the *Ancient Mariner*. What the experiment confirmed that rainy afternoon in the botanical gardens was the wonderful willingness of the general public to take part in eccentric activities requested of them by a complete stranger. It was less successful, as they lent in one by one to sniff a blossom, in eliciting interesting descriptions of the fragrance, illustrating our general lack of respect, and our paucity of vocabulary, when it comes to the matter of smell. It was 'nice', it was 'pleasant', it was 'subtle', it was 'not like a rose at all'. At the more imaginative end it was 'citrus', it was 'sandalwood', and it was 'not quite nutmeg'. The one thing we all agreed on was that it was not what we thought of as 'musky'.

* * *

If it was not in the musk rose, might I find the fragrance among the British wild flowers that had 'musk' in their name? There were four of them that I could think of: the musk mallow, the musk orchid, the musk thistle and the musk storks-bill. There were, too, numerous plants which had 'musk' in one or other of their many country names; the rather unlikely sea aster, for example, which was once known as 'musk buttons'. Then there was the tiny moschatel, whose name means 'little musk', and it was this, by chance, that I was to come across first.

It was a morning in March and I was on an expedition with my friend Dave Bangs. Dave combines his work as a gardener with his radical politics and his ceaseless campaigning on access issues and the protection of open spaces. He has, too, a painstakingly acquired and impressively detailed knowledge of the wildlife of Sussex: where to find the lemon-scented fern, the spawning grounds of the brook trout, the nests of wood ants or the wintering sites of Bewick's swans. I had been out with him on many exciting explorations but on this occasion we had failed in our mission to hear the singing of wood larks, despite getting up at 3.30 in the morning and visiting several different Sussex heaths. Heading back towards Brighton, Dave asked if we could make a final stop by the eighteenth-century Ambersham Bridge near Graffham, for the obscure purpose of locating a plant called the alternate-leaved golden saxifrage. This was scarce in the county, he assured me, adding with his usual precision, that its only sites were on the 'Bargate beds of the Western Rother'. The Bargate beds were not a particular passion of mine, but he had put me up for the night and driven me round since long before dawn, so the only correct course was to accede. We got out of the car and walked, or more accurately splashed, around in a tiny wood of neglected hazel coppice, damp birch and alder; the still-bare trees tall and thin and creaking in the breeze like an unoiled gate. Beneath them rose a tiny spring, spreading out across the woodland floor, the dampness ensuring the understorey was lush and thick with herbage. There were seedy spikes of dog's mercury, tangles of hemlock water dropwort, large butter-yellow blooms of kingcup and scatterings of the lovely, lilac-tinted lady's smock. Robins trilled, in their usual melancholic manner, and an early chiffchaff called with the persistence of a dripping tap. And we did find the alternate-leaved golden saxifrage, together with its commoner relative the 'opposite-leaved' species, the 'gold' of their tiny flowers being of the unburnished kind, like something fresh from the mine that had yet to be polished and prepared. And there, on a drier bank, was a patch of the equally diminutive moschatel. Its scientific name *Adoxa moschatellina* indicates its lowly status, for *adoxa* comes from the Greek word for 'inglorious'. But it is attractive enough, in its own unshowy way, with its mid-green, three-lobed leaves resembling those of a miniature flat-leaved parsley. It is also distinctly curious, its country names of fairies' clock and town-hall clock reflecting the unusual structure of its flowers. Borne on fine, pale stems they form a head the shape of a dice cube, with a flower flat on every face, except the underside. The moschatel thus looks in every direction simultaneously, even directly into the sky. The terse descriptions that usually accompany the lavish illustrations of my beloved Keble Martin consist almost entirely of botanical terms: hispid, scarious, pinnatifid, corymbose. Only on the rarest of occasions does he drop his guard, and the moschatel is one of them. Unable to resist his priestly vocation, he adds to its description the unscientific sentence, 'A symbol of Christian watchfulness'.

It was these flowers that we knelt over as the water seeped into our jeans and the damp woodland soil stained our knees. Farm tractors and the ubiquitous four-by-fours trundled past on the nearby road and if any of their occupants did happen to look in our direction, we would have presented a curious sight. We smelt the flowers, their golden anthers shining out from among their green petals. We smelt the leaves. We rooted in the soil and tried to smell the tubers. We sniffed our best, trying to detect what I had read was a 'faint musky smell . . . particularly noticeable on warm, damp, spring days'. Roy Genders, in his book on scented wild plants, had given an even more specific description: 'Upon approaching,' he says, 'the whole plant emits a muscat scent, to some people resembling almonds, to others elder blossom'. The best we could come up with was the smell of herbage and of humus, and a faint tang of damp dog's hair. It was interesting, but it was definitely not erotic.

A couple of months later I was driving back home from the City of London Cemetery, through some of the road-fringed fragments of Epping Forest which hang on in urban east London. Suddenly, and almost subconsciously, I spotted some stems of musk mallow growing on the roadside. I brought the car indecorously to a halt as soon as it was semi-safe to do so. Driving and the observation of wildlife are not the most comfortable of companions. I culled one of the flowering stems, for which I ask forgiveness, took it home and stood it in a glass of water. It is a pretty plant, its lacy leaves so deeply divided that they are more air than leaf, its wide-open flowers with spaced purple-pink petals and a projecting furry anther. It is said of the musk mallow that its fragrance grows in intensity once it is brought into a warm room, and so it proved to be with mine. With this plant I did seem to be getting closer to my goal, for it produced a slightly animal tone overlain with higher notes that were not rose or lily but lightly floral – sweet pea perhaps. Over the next few days it seemed to become more intense, and insisting, as I did, that every visitor sniff it, produced descriptions of ancient perfume bottles, laundered clothes, old houses and several visits to a grandmother.

Dave had also arranged for me to visit a Sussex downland site where the musk orchid and the musk thistle grew together. But the window of opportunity was short and with work and family commitments I missed it. I was never to test whether, as the sixteenth-century herbalist John Gerard put it, this nodding thistle really does have a 'most pleasant sweete smell, striving with the savour of musk', or whether the orchid has, as Roy Genders suggests, a 'heavy scent reminiscent of honey'.

I was left with just one more plant I could find that season. The musk storksbill, according to the books, is a plant of the West Country, found normally close to the sea. In an effort to pinpoint it more exactly, I invited another friend, the botanist John Swindells, to lunch at my local café, without disclosing my reason to him in advance. When I arrived he was already sitting at one of the rather eccentric assortment of tables that characterise the place. Arriving early, he told me, he had decided to take a walk around to see if he could find anything of interest. And he had; a plant, he told me, called musk storksbill, growing 'just outside the café door'. It was a wonderfully fortuitous moment and one that I like to take as an example of synchronicity.

Once John had introduced me, I was to find it all over London. It is a plant that hugs the ground, its hairy stems stout and sprawling, its leaves divided into paired and jaggedly toothed segments. Its rose-pink flowers are formed in clusters of six or seven and they set fruit in long tapering pods gathered closely together as if in huddled conversation. These are the eponymous 'storksbills', though to me they collectively resembled some strange surgical instrument rather than an ornithological appendage. There is even a touch of Edward Scissorhands about them.

I was finding the plant on lawns, at the base of fences, in pavement cracks, and one particularly fine specimen was growing less than a hundred yards from my own front door. And yet the 1983 edition of the *Flora of the London Area* lists the musk storksbill as occurring in just three sites in the whole of Greater London. Why was this plant, once so restricted in distribution, now such a common street-weed? The answer, according to Rodney Burton, writing in *The London Naturalist* in 2005, is most probably dogs. In its West Country heyday its sticky seeds would have caught in the wool of sheep, inadvertently transferring the plant from place to place. Now, the suggestion was, they were catching in the hair of holiday-makers' dogs, leading to its spread to new parts, and over much greater distances.

It seems I had the canine population of Poplar to thank for the fact that I could examine so many specimens of the plant over the next few weeks. Leading a local walk during that period, I persuaded the whole party to stop and sniff it. In Patrick

Suskind's historical fantasy novel *Perfume*, which explores the relationship between scents and emotions, the murderous perfumer creates a fragrance which reduces a whole crowd to an immediate and indiscriminate orgy. I was not, necessarily, looking for such an intense reaction but it would be interesting to see what descriptions people came up with. I repeated the exercise with other friends and family, and the responses were rather more productive than my musk rose experience. They ranged from 'herbal tea' to 'leather', to the 'far-away scent of cloves'. According to one description, the smell was reminiscent of a 'damp, green dell, thick with ferns, in a Scottish forest, and with a beam of sunlight appearing between the trees'. But that was my brother-in-law and I believe he'd had a whisky or two. Some people suggested floral, fresh or fruity tones. It had a hint of lime or strawberry, I was told, 'a slight sweetness' or 'the smell of mown grass'. But other descriptions were earthier, more bodily, and sometimes unsettling. It smelt of potatoes, of mushrooms, of dog or human urine, of sickness or 'the sick room' or 'the breath of someone who is distinctly unwell'. Clearly, as Arthur J. Stace had implied, the smell of musk was not to everyone's taste.

The plant, I was soon to discover, also had other aspects to its history. In his sixteenth-century Herbal, John Gerard describes it as planted in gardens 'for the sweet smell that the whole plant is possessed with'. And, returning to the library of the London Natural History Museum, I came across a half-finished copy of *The Plymouth and Devonport Flora*, privately published by George Banks in 1831. The 'Musk storksbill, Heron's-bill, Muscovy or Pick-needle', he wrote 'is much esteemed for its musk-like fragrance and generally propagated with other garden annuals, nor indeed are any of the beautiful nosegays which deck our market stalls, considered perfect without a branch of Muscovy or Pick-needle'. The musk storksbill, it seems, was being sold on stalls in Devon and Cornwall at about the same time as the musk plant itself was becoming popular in the markets of Liverpool and London.

And what did I make of it personally? The strength of it seemed to vary from plant to plant, but at its strongest there was a distinct mingling of sweetness and sweat. It had the animal odours of a badger or fox or the swept bedding of a goat's pen, and yet at the same time the fragrance of honey. Like many smells, it provoked strong memories: of childhood visits to the animal houses at London Zoo, or of dancing in a marquee at a music festival with the odour of massed bodies mixing with the fragrance of pounded grass. In this ability to provoke memories, in its mixing of tones, in its one-time use as a garden plant and its place in West Country posies, the smell of the musk storksbill was, I thought, as close as I was going to get to experiencing the lost fragrance of the flowering musk.

* * *

None of this had got me any closer to understanding why that smell had actually been lost. If there was a significant year in the story it was, it turned out, not so much 1913 as 1930. It was in that year that a single speech was to bring the plant back to the forefront of public attention and to spark a new search for an explanation. It is hard in the age of multiple media platforms to recapture the impact that could once be made by public speeches. They could form the heart of political campaigns and important social debates. They could attract huge audiences and be reprinted verbatim in the national press. And they could be delivered with oratorical flights that are lost in the days of Trumpisms and tweets.

With its ungainly title of 'Recent Developments and Present-Day Problems in Taxonomic and Economic Botany', this particular speech does not seem such an immediate candidate for controversy. It was delivered by Sir Arthur Hill, Director of the Royal Botanical Gardens, as his presidential address to the

British Association (Section K: Botany), and dealt with what he described as 'some imperial problems'. These included difficulties with the fruiting of avocados in Bermuda, the immunity to wither tip disease of limes in Dominica and the production of strains of banana immune to Panama disease. It touched upon the difficulty of defining a species and bemoaned the tendency of young scientists, eager for recognition, to continually identify new plants when often what they were describing were simply different growth forms. And in the middle of all this, it referred 'in passing, to that peculiar and elusive subject, the loss of scent of the common musk'. What was different here, however, was its depiction, for the first time, as a worldwide phenomenon. According to no less an authority than the Director of the Royal Botanic Gardens, the plant had not only lost its scent in Britain but across the world, including its native home in British Columbia.

Buried though it was in a longer speech, it was this short section that caught the attention of the press. It was reported in the *Observer*, *The Times*, the *Daily News and Chronicle* and *The Spectator*. It featured in the *Bristol Times and Mirror*, the *Newcastle Journal*, the *Glasgow Daily Record* and the *Bolton Evening News*, and no doubt in other regional papers that escaped my researches. The horticultural journals were equally excited. 'Here is a mystery', ran an article in the *Gardeners' Chronicle*, 'as straightforward as that in the infinitude of modern detective stories which provide you with the corpse in the first chapter and keep you more or less hot on the scent until the last.' It was this speech that started the stream of correspondence in the journals, which came to include confirmatory accounts of the loss of fragrance from New Zealand, the Sierra Nevada Mountains of California and from several other parts of the globe.

There is, in the archives at Kew, a small brown folder containing Hill's handwritten notes for the speech. With it is a carbon copy of some of the original typescript which Sir Arthur had sent to an unnamed friend, and although the accompanying letter is not available, we do have the unsigned response. It is written on the headed paper of the Adelphi Hotel at St Leonards-on-Sea and dated 16 September 1930:

> *Dear Hill,*
>   *Thanks, many, for your interesting notes on the Common Musk and extract from your address in Bristol.*
>   *It is a most puzzling riddle and if the loss of scent is due to the dominant hybrid plant why should the conquest be so widespread?*
>   *We are here for two or three weeks to escape spring cleaning and I am glad to say that my wife is none the worse for the discomforts of hotel life . . .*

Apart from its glimpse into the sufferings of the early twentieth-century well-to-do, the letter demonstrates that attention was now turning to the possible causes of the phenomenon and it was from this time that theories began to proliferate. They included chemicalised bees, the spread of the petrol engine and the impacts of influenza. They varied, in short, from the preposterous to the profound.

One of the strangest among them was the suggestion that the plant had never actually had a smell. It was all a kind of 'Emperor's new clothes' syndrome, an idea that had been passed from person to person without anyone properly checking it out at all. It was a theory that was to gain more adherents as the years passed and as the generation that could actually remember the fragrance died out. Discussing the story in the 1990s with a Professor of Botany from Exeter University, it was still one of the avenues he suggested I should investigate. 'Are there any Victorian books which actually mention the smell?' he asked me. Well, it turned out, there were, and many of them. In addition to which it would be hard to overlook the evidence of the nursery-men who had sold the plant, the botanists who had studied it or the women of the Edinburgh asylums. And why would Lindley and Douglas have named it *'moschatus'* in the first place, if it was scentless? But this account is sustained by the particular, and notable, fact that Douglas had failed to make any mention of a fragrance in his journals.

A theory that came from a completely different direction was that the plant did have a smell, and continues to have it, but that humans have lost the power to detect it. One version of this came from a Dr F. Dawtrey Dewitt who wrote to *The Times* shortly after the Hill speech:

> *Our sense of smell has become much less acute than that of primitive races and vastly inferior to that of lower animals. Is it not possible that the frequent and universal epidemics of influenza accompanied by nasal catarrh since 1890 have quickened the rate at which civilised man is losing his sense of smell? It would be interesting to note whether members of primitive races or very young members of our own agree that the musk plant has no scent.*

The imperial language aside, there were others who espoused the same idea, whilst suggesting different causes for this possible olfactory atrophication. It was, rather, suggested a correspondent to the *Gardeners' Chronicle*, the rapid spread of the motor-car and its attendant 'stink' that had deadened our sense of smell, an idea we could test 'by taking (the plant) to a lonely island in the Pacific where the air has not yet been polluted by the deadly fumes of the petrol engine'.

There may, I suspect, be some truth in the idea that our sense of smell is declining, or at least that we are making less use of it, though the phenomenon is cultural rather than biological. There is a clue in those condescending references to 'primitive races' and 'lower animals', as well as to children. It is as if, in some supposed coming of age, we have laid aside our sense of smell as too closely associated with an earlier state; a previous stage of human development or our own childhood. Smell is one of the proximity senses, so intimate as to be almost distasteful, and we feel we have to grow out of it. A mark of this reaction is the strange fact that to tell someone they 'smell', without the need for any qualifying adjective, is not a neutral statement of fact but the most personal of insults. And the fear of that smell, and the attempt to conceal it, has generated a billion-dollar industry. Notwithstanding the flu and the petrol fumes, there is a need, if we are to reconnect with the natural world, to reacquaint ourselves with its fragrances: with the smell of plants and rocks and trees and soil, with the balsam-scented buds of the poplar, the sharp tang of the tansy, the pollen-rich fragrance of the dandelion, the sickly-sour odour of black horehound leaves, the fresh aroma of rain on parched earth. All of these things are still available, if only we take the time and effort; and the fact they are still available means that the explanation for the musk plant's loss of fragrance must lie elsewhere.

Another group of theories sought explanations in the interaction between the musk plant and other organisms. One suggestion was that bees fed on 'medicated candy', presumably then a widespread practice in bee-keeping, had inoculated the plant with chemicals that subsequently inhibited its smell. More subtle, perhaps, was the idea that the scent had never been a product of the plant itself but had come from the presence of minute organisms with which it lived in symbiosis. It was these that had been suddenly and universally lost, a suggestion which dealt with one intractable problem by simply raising another. Fragrance in plants can serve the seemingly contradictory functions of attracting insects or deterring them, and both these functions were covered in theories propounded at this time. W.H. Cope, from the Field Naturalists' Club, argued that the fragrance of the musk had evolved to attract a specific group of pollinators and that since these insects were absent from this country, the plant in cultivation had shed the smell along with the need for it. Arguing from the other angle, Dr Eugene Charabot suggested that the fragrance had evolved to prevent the attacks of leaf-eating insects, as with the strongly smelling herbs of the Mediterranean maquis, and it was the absence of these insects in cultivation that had brought about a similar result. An interesting variant on this came from a letter in *Nature* which suggested that the specimens collected by Douglas had come from higher, drier ground than was the usual habitat for the plant and were of a specific form that had developed scent glands for protection against grazing animals. It was a process, the correspondent suggested, similar to the way that the holly develops prickles only on its lowest leaves, the ones that are most susceptible to grazing. The theories are not unreasonable even though they suggest a remarkably rapid rate of evolution. What would seem to undermine them is the worldwide nature of the phenomenon. It was not just the introduced plants in British cultivation that were affected; it had happened globally, including to those in the wild.

While botanists, biologists, beekeepers and, indeed, the general public, were having their say, the horticulturalists had not remained silent and were producing theories of their own. Some considerable argument took place over the introduction, in 1877, of a new, hybrid version of the plant. A cross between *Mimulus moschatus* and the related species *Mimulus luteus*, it was named Harrison's musk after the nurseryman who had developed it. It was larger-flowered, scentless and, like many hybrids, very robust. The Harrison's musk, it was argued by some, was displacing the original musk plant, or, by a process of unchecked cross-fertilisation, was depriving it of its smell. Others, including the Head Gardener at the famous Nymans Gardens, responded indignantly, that the two were quite distinct in appearance and that he would have noticed if one was supplanting the other. Moreover, plants were losing their smell in areas far removed from any source of Harrison's musk. Another notable figure to join the debate was Sir William Beach Thomas. Thomas had been well known as a war correspondent and was now a writer on gardening and wildlife, as well as being an early champion of the National Parks. His theory was that the problem actually lay in the lack of pollination, arguing in *The Spectator* that the plant had been reproduced too long from cuttings with the result that 'vitality of the stock wilted and along with the energy, vanished the scent'. Among others, attention turned to the fact that the plant grew in both hairy and hairless forms. The fragrance, they suggested, was found only in the hairy form and that when this was lost, the smell went with it. Though it seems an argument that is more about mechanism than about cause, it was still being put forward in books as late as the 1970s, despite the fact that some sixty years earlier the Jodrell Laboratory at Kew had tested both hairy and hairless plants and determined that both could be equally fragrant.

It was amidst this swirl of competing theories that the Scientific Committee of what was now the Royal Horticultural Society convened a special meeting on 11 September 1934. There is a typed copy of the minutes among the archives at the Lindley Library and they bear the signature of the committee's Honorary Secretary, E.H. Bowles. A botanist, entomologist, artist and writer, he was another of the giants of our horticultural history and his gardens at Myddelton House have now been restored and opened to the public. Like Frances Jane Hope he had put them to philanthropic use, though in a rather different way. He worked extensively with local boys' organisations and at the weekends laid on activities for them in the gardens, including fishing, football, cricket and, in winter, skating on the pond. Many of these 'Bowles boys' had helped him, too, with tasks in the garden, digging, planting and clearing the pond of weed. I was excited to see his signature, but even more excited to read the minutes to which it was attached, for they turned on its head my way of looking at the problem.

The minutes begin by recapitulating the discovery of the plant and its subsequent cultivation at the Society's gardens in Chiswick, noting, again, that Douglas does not mention its fragrance in his journals. The plant, they say, is common in British Columbia and around the western parts of the USA and has now run wild in many places in England, including the streams around Wisley, the site of the Society's new gardens. All quite straightforward, but then comes the suggestion that the scented plant grown from seed at Chiswick was not actually the normal form of the plant but an aberration. As long as this plant was propagated vegetatively, all of its 'offspring', being genetically identical, would be fragrant. But this would not be the case if new stock was introduced or if new plants were grown from seed. 'The musk', the minutes concluded, 'has not lost its scent, but the scented form has been lost'. Little attention seems to have been paid, at the time, to this meeting or to its dramatic suggestion. Unlike the Hill speech, it had gone unreported in the press and unremarked in any contemporary journal. Yet its conclusion was profound. The mystery, I now realised, was not that the musk had stopped smelling; it was that the musk had ever smelt at all.

From this point it was relatively easy to piece together the story. David Douglas collects specimens of the musk plant in British Columbia. The reason he doesn't mention in his journal that they were scented is because they weren't. He also collects seeds of the plant and these are grown on by Lindley in the Society's garden. It is from these seeds that an unusual and aberrant form arises. We have the further knowledge today that the taxonomy of the whole *Mimulus* genus is complex and that other species, used in genetic studies, have been described as 'evolution happening before our eyes'. Up to now, *Mimulus moschatus* has been regarded as the stable species among them, but perhaps it is less stable than was previously thought. Here then, in the Chiswick garden, a rare recessive gene comes to the fore, and a scented plant appears. Who, then, is to realise that this is the exception rather than the rule? Complaining of the poor quality of the specimens Douglas had brought back with him, Lindley uses the Chiswick plant to formulate his scientific description and the *Mimulus moschatus* is now officially scented. It is the accepted, international description of the plant that will be repeated in Floras around the world. As for the Douglas notes, these were compiled any time up to two years after his return. It is entirely possible, therefore, that he, too, uses the Chiswick plant to refresh his memory and simply assumes that he had previously overlooked the fragrance.

It is this, too, that is the source of the entire original stock of cultivated plants, propagated over and over again from cuttings. As long as it is reproduced vegetatively it replicates all the features of the 'parent' plant, including its aroma. It becomes the basis of the whole horticultural trade. But *Mimulus moschatus* is, as we know, a short-lived perennial. In gardens like that of Mrs Gardner, where the plant 'came up year after year', the original plants are dying out and what is replacing them are self-sown seedlings, lacking the recessive gene and inevitably reverting to the original, scentless type. Given the popularity of the plant and its huge sales, it is likely that stocks are replenished with new imported supplies, or with seeds collected from other plants. All of these will be scentless. It takes a couple of decades but gradually it is this new material that is taking over. Meanwhile the scented form has been exported all over the world, even to the gold diggers of Otago, and even back to its native areas of north-west America. Familiar with this form, it is only later that people begin to notice that the wild plant is unscented and assume that this is a recent phenomenon, thus bolstering the claims of a sudden worldwide loss of fragrance. Ironically, it seems that there is an element of truth in several of the theories. Perhaps, as Beech suggested, the stock generated over and again from cuttings was 'losing its vitality'. Perhaps the scented form does only arise in certain habitats or conditions. There was even something in the idea that the plant had never smelt at all, although not quite in the way its proponents had suggested. And so with a possible solution to one mystery, came the unveiling of another. Where, why and when does *Mimulus moschatus* produce its rare scented form? And will it ever happen again?

For several years after the 1930 speech, reports were being received of wild musk plants that had produced a scent: from Marlborough in New Zealand, from Millstream in the very south of British Columbia and from Texada Island further north. As interest in the phenomenon died down, so the reports dried up. But it is possibly still out there. Or the recessive gene will resurface and we will once again have a scented specimen from which we can take cuttings and propagate. And then, once again, we could experience this plant of Douglas and Lindley, of Hooker and Hope, with its enigmatic, exotic and erotic fragrance; the plant that was once sold on our streets, handed out in posies and graced the lives of thousands of ordinary people.

2

Death and Life in the Churchyard

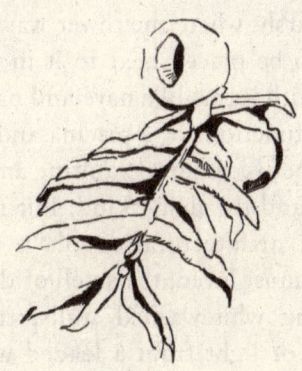

I had a rather unusual interest as a child, and one that I would not have admitted to in front of friends. It seemed, even then, a rather fusty pursuit, and my siblings certainly seemed to think so. But there it was, I loved to visit old churches. Every year we would pack into my father's black Ford Popular, a breed of car then familiarly known as a 'sit up and beg', and set off on the same family holiday. And every year during those long seaside summers, staying with a beloved great aunt in Sandgate, I would demand that we undertook a tour of the Kentish village churches. Contrary views would be expressed, sometimes forcefully, but just once in a while I would carry the day, usually because it was raining and there was little else to do. So back we would get, into the upright little car with the number plate NYO 714, a fact I mention only because I have never been able to recall the registration of any other car with which I have since been associated.

In this rather uncomfortable and over-crowded fashion, we would follow the narrow high-hedged country lanes, my father hooting at every bend, my mother perpetually telling him to drive more slowly. And in this way I got to know places like Lenham, with its fourteenth-century wall paintings, Hythe, with its rows of skulls in the crypt, Dymchurch with its stories of smuggled brandy in buried coffins, and Brookland in the Romney Marsh, where the tower was too heavy for the church and had to be placed next to it in the churchyard. I became familiar with terms like nave and narthex and ambulatory. I knew the functions of a piscina and a squint. I could tell the difference between barrel vaulting and a wagon roof, a standing buttress and the flying kind. But more than any of these historical or architectural details it was, I think, the atmosphere that I most loved; the smell of damp hymn books and of the flaking whitewashed walls; the semi-darkness pierced by beams of light from a leaded window and filled with dancing dust motes; the strange deep silence through which you could hear, if you listened intently, the sound of the universe expanding.

That same atmosphere extended to the churchyard. From the lychgate, with its coffin rest, I would wander up to the church cross or the timbered porch, passing among the table tombs, the railed vaults, the stone angels and the leaning slabs, looking out, with an anticipatory shiver, for those older stones which bore that sinister symbol of mortality, the skull. Just as sepulchral were the yews. They stood, in supervisory fashion, beside the gate, or lined the path that encircled the church, or extended their boughs to shade the grassy rows of graves. They were not just dark in themselves, it seemed to me, they drew darkness towards them. They were such an essential part of the scene that I never actually questioned the connection. They were an inevitability, something inseparable from the rest of the place. It was only later that I began to puzzle over their presence. Why were there so many yew trees in churchyards? And why were there so many churchyards with yews? And what was the origin of this seemingly inseparable relationship? It was, I found, a question that had long been asked, and one that had long remained unresolved.

The yew is a tree of multiple contradictions. It is a conifer, but unlike any other member of its family. It reaches a great age, but one that is almost impossible to determine. It is simultaneously sacred and secular; a part of our spiritual topography but also our military history. It is poisonous but also protective. It is associated with Christian theology but also with pre-Christian belief. It is connected with death but ascribed with life-giving powers. And of all its many mysteries, none has given rise to as much debate as this long-standing relationship with the churchyard. Mentioned by Giraldus Cambrensis as early as 1187, it has become part of our literature. It is discussed by the learned Thomas Browne in his *Hydriotaphia* of 1658, and by the diarist John Evelyn in *Sylva* in 1664. It is debated by the famous Gilbert White in his 1789 *Antiquities of Selborne* and, around the same time, it is argued over in the pages of the world's first 'magazine'. When the physician John Lowe publishes, in 1897, what is probably the very first book devoted entirely to the yew, it contains a whole chapter on the subject. And it has been debated, sometimes heatedly, in books, articles, learned papers and web pages ever since.

Disagreement extends even to the number of churchyards with yews and I have seen estimates ranging from 500 to 'more than 2,000'. The Ancient Yew Group, which has done so much to document the tree, lists 1,350 churchyards in England and Wales that contain 'ancient' or 'veteran' trees. Those in Scotland add more, and though they are fewer in number they include some of great significance, including one that is reputedly the oldest tree in Europe. But the story of the churchyard yew is not just about these ancient, large or exceptional specimens, for this is a living and continuing tradition and the more to be treasured for that. People have planted yews next to churches across many centuries, and they still do so today, for as recently as 1999 the 'Millennium Yew Project' claimed to have distributed over 7,000 yew seedlings to British churches.

I broke off, in the course of writing this, to walk, with our dog, from my home in Poplar up to the neighbouring parish of Bow. There, in the centre of the busy main road, sits St Mary's, the traffic splitting and rushing around both sides of the old stone church, like a mountain river breaking around a rocky island. Squeezed between the carriageways, the churchyard is long and narrow, and completely overhung by lines of tall London planes. Their arching branches meet in the middle so that entering the churchyard gives the impression of walking into a great leafy vault. And even in this shaded, grimy, fume-laden place, young yews have been planted, and line the stone path leading to the eccentric tower and the main church door. Take these recent plantings, take the young trees and the middle-aged alongside the venerable ancients, and the total number of churchyard yews must rise well above the more conservative estimates, even if only a few of those Millennium plantings are to survive.

What is just as significant, as far as our most notable yews are concerned, is that very few of them can be found anywhere outside of church grounds. The figures from the Ancient Yew Group suggest that 67% of all the older yew trees in the country are to be found in churchyards, and for all those with 'large girth' the figure rises to 73%. It is a remarkable fact, and one that requires some reflection, that less than one third of our oldest and most significant yews can be found anywhere at all outside of this one habitat. As Richard Mabey was moved to comment in his *Flora Britannica*: 'I do not know of any similarly exclusive relationship between places of worship and a single tree species existing anywhere else in the Western world.' It was this striking phenomenon which, with resounding echoes of my childhood interest, I had set out to investigate.

* * *

One of the early debates on this puzzling phenomenon took place in the pages of *The Gentleman's Magazine*. Established in 1731, it was not, as my wife suggested when she first saw me consulting it online, an early version of *Playboy*. Instead for nearly 200 years this rather earnest and erudite digest, with contributors including Samuel Johnson and Jonathan Swift, discussed science, politics, history, literature and any other topic deemed to be of interest to the educated mind. It was also the first publication to call itself a 'magazine', a name derived from the French word for storehouse. Its articles took the form of a correspondence addressed to a 'Mr Urban', the pen name of the magazine's editor, and from 1779, continuing through the 1780s, these included a series of contributions on the conundrum of the churchyard yew. Disagreeing with each other with more or less polite disdain, the writers offered a series of conflicting explanations. These theories, set out over 200 years ago, remain pretty much the same as those under discussion today. The first of them, and the one that came to set off a particularly long and intense debate, was the suggestion that the origin of the association lay in the long history of the longbow. First put forward in these pages in 1779, it returns several times and reappears in 1830 in a particularly patriotic and folkish fashion. As one anonymous correspondent wrote:

> When invasion or sudden attack was apprehended to the churchyard might simultaneously resort the inhabitants of every parish, and there speedily supply themselves with weapons, as from a common armoury. The lopping of branches for such a purpose would not come within the interdict 'Ne Rector arbores in cemetris prosternat', [Let not the Rector lay low the

trees in cemeteries] *because no tree, perhaps, sustains so little injury by lopping as the yew. Lopped, moreover, under such circumstances, the severing of some of its branches would be done by the natives with care, and even with veneration; considering it almost as a sacred beneficent guardian, that was at every future crisis to yield them and their children a further supply.*

It is a rather Dad's Army view of the Middle Ages and one that is highly inaccurate.

If the wheel, gunpowder, penicillin and the atom bomb are among those inventions generally recognised as having shaped the course of human history, there are others that are less readily acknowledged. The stirrup is one of them, and so is the bow. Here was a device, developed in many different cultures, which could propel a projectile faster, with more deadly force and over a greater distance. By making hunting both safer and more efficient, its contribution to the survival of our species was immeasurable. And, as is the case with almost every technological advance, it also made warfare more deadly. The association of the bow with the yew tree can hardly be older. In September 1991 two German tourists, walking in the Ötztal Alps, came upon a mummified body in the ice. 'Ötzi the Iceman', as he came to be known, was eventually dated to 3300 BC and among his possessions were a quiver of arrows and an unfinished longbow made of yew. If Ötzi lived by the bow, he died by it too, for scientists determined the cause of his death as an arrowhead found deeply buried in his left shoulder. Though one of the most studied, the Ötzi find is not the earliest yew bow so far discovered and there is a British bow that predates it. In the same year that the German tourists made their alpine discovery, a hill walker in the Scottish borders east of Tweedsmuir, came across a broken longbow in a peat bog known as Rotten Bottom. He kept it in his garage for some time before deciding to consult an expert, who dated it to sometime between 4040 and 3640 BC. From this Middle Neolithic period through to the Anglo-Saxons, the bow was in continuous use in these islands, primarily for hunting. All that was to change, however, with the Norman invasion and their strategic military use of the bow.

A big advance in the technology of the longbow had been made somewhere in Scandinavia, in the first three centuries AD. Here it was discovered that different layers of the yew tree could be combined to a particularly powerful effect. The heartwood of the yew resists compression, the sapwood is highly elastic. Put these two together, the heartwood on the inner side or 'belly' of the bow, the sapwood on the outer, and you have a weapon of much greater tensile strength. Moreover the two didn't need to be artificially joined; they could be cut as a natural composite from just beneath the bark of the tree. The result was a Viking bow that could now fire an arrow, with great force, as far as 660 feet. This was no longer a bow that was primarily for hunting; it was a combat bow. The Normans, or 'Northmen', were themselves descendants of Viking settlers and it was with this bow that they conducted their English invasion; an invasion in which Harold II became the first English king to be killed by an arrow.

The history of arms, and of the arms trade in particular, is full of ironies. It is perhaps one of them that a French invasion brought to England the weapon that was to be so effectively turned against France some two centuries later. And ironic, too, that a weapon arising in Scandinavia, arriving via France, and very soon to be mastered by the Welsh, had, by the Middle Ages come to be seen as synonymous not just with English military prowess, but with the national character. It had, by then, undergone further technical changes, being now nearly twice as long as the Norman bows depicted in the Bayeux Tapestry. The range of the bow had now reached 980 feet, but mastering such a large and unwieldy instrument was not easy, requiring physical strength and up to ten years of regular practice. But that practice had become a national discipline. As John Lowe put it in his 1897 book, *The Yew Trees of Great Britain and Ireland*, 'from the time of the Conquest onwards the [long] bow became the national arm, in the use of which the English acquired more proficiency than any other nation'. English longbow archers were used by Henry II in his invasion of Ireland and accompanied Richard II on the crusades. Edward I, in the late thirteenth century, incorporated archers into the English army, having seen the longbow used very effectively against him by the Welsh. He ensured a reservoir of more or less willing recruits by passing an edict that required every able-bodied man, with a few exceptions such as clerics, to own a bow, to keep it in good order and to engage in regular practice. It was the first of many such edicts over the coming centuries and came to regulate almost every aspect of its provision and use. It was used with terrible efficiency in the repeated wars with the Scots but, more than any other period, it is with the set-piece battles of the Hundred Years War that the English archer and his longbow have come to be most associated. From 1337 to 1453, five generations of English kings, from two rival dynasties, threw thousands of lives into the defence of their hereditary French possession. Among the major battles of the period, Agincourt, dramatised, and romanticised, by Shakespeare, has come to be the best known. But the battle which was the first and most decisive, and which established the supremacy of the longbow, was Crécy.

In 1337 King Philip II of France repossessed the territories held in his country by the English monarchy. Some eight years later, Edward III responded by landing a huge force on the French coast and marching with them through Normandy. To finally meet up with the French he chose a location at Crécy-en-Ponthieu, arranging his battalions at the top of a slight rise, his infantrymen flanked by contingents of English and Welsh archers. More bowmen hid in a little wood or lay down to conceal themselves in the wheat fields. Having been there since dawn they were well rested by the time the French arrived, with an army that now greatly outnumbered the English, in some accounts by as much as ten to one. They were led by a force of Genoese mercenaries wielding crossbows. These, they planned, would 'soften up' the English infantry and disorder their formations, allowing the French men-at-arms to break through, a tactic that had been used effectively elsewhere. As they took up their position in the van of the French forces it was apparent that this really would be a test of the two bows.

The crossbow had a number of advantages. It did not depend on physical strength and required less training. It could be put into shooting position and held there till needed. And its winding mechanism produced a greater armour-piercing impact over a longer range. It had, however, one great disadvantage. It was slower. In the time it took the longbow to loose up to ten arrows, the crossbow could manage just two. And it had another downside that was just about to be revealed. As the two forces formed up, there was a sudden, sharp shower of rain. The English bowmen quickly removed their bowstrings and placed them under their helmets – a practice that is said to have given rise to the expression of keeping something 'under your hat'. The crossbow strings could not be removed, with the result that they slackened and became less effective. In the archery exchange that followed the mercenaries were quickly routed. Having, for some reason, left their shields in the baggage train at the rear, they began a hasty retreat through the French lines. Deeming them traitors, the French knights hacked and slashed at them as they passed, killing more of them in this way than had died from the English arrows. The main French forces now pressed forward, unfurling their sacred oriflamme, which indicated that no prisoners would be taken or quarter given. They were hampered by the retreating Genoese, by the muddy slope and by the pits which the English had dug. The storm of arrows that descended on them numbered as many as 70,000 a minute, 'flying in the air as thick as snow, with a terrible noise like a tempestuous wind', wrote one eyewitness. Few of the French made it through to the English lines and those that did were cut down in the hand-to-hand fighting that ensued, fighting described at the time as 'murderous, cruel and without pity'. Throughout the afternoon, the French repeated their increasingly pointless charges, now additionally impeded by the piles of dead and dying men and horses. Between each attack the English archers would rush forward to retrieve some of their arrows, then wheel back into position ready to loose them again on the next attempted advance. In the course of this battle, according to modern historians, as many as half a million arrows were discharged. By midnight, some twelve to fifteen charges later, it was over. The exact number of casualties is much debated but there is no doubt that it was highly asymmetric; the English dead estimated at around three hundred, the French anywhere between twelve and twenty thousand.

* * *

Crécy helped establish the longbow as the English national weapon and it was long to remain that way. Its last significant use in a major engagement was at Flodden Field in 1513 and by that time it had survived the introduction of the gun by over fifty years. When Elizabeth I began to replace bows with matchlocks in the English army it was not for military efficiency, for the bow was still the faster and more accurate weapon; rather it was to do with the recurring, and now worsening, problem of securing a sufficient supply of yew wood. It is this persistent shortage that remains the rationale for the supposed churchyard connection. In order to maintain the bowyer's supplies, it is argued, yews were planted on church ground which, being both walled and sacred, was doubly protected against grazing livestock, casual lopping and other depredations. It is a theory that is supported by a number of contemporary churchwarden accounts, but by very little else.

The long years of the longbow's supremacy were marked by dozens of royal edicts governing even the minutest details of its production and use. But as 'AB' points out to Mr Urban, in one of his contributions to *The Gentleman's Magazine*, nowhere do they mention the planting or use of churchyards. There is, instead, at least one piece of evidence to the contrary for, when Henry V was preparing his French campaign, the campaign that would eventually lead to Agincourt, he empowered his bowyer to enter private land in order to requisition yew wood. From this, however, church land was specifically exempted. The churchyard, it seems, was the preserver not the producer of yew wood, and this despite the fact that demand for the wood was becoming almost insatiable. Every bow required a six-foot length, straight, free from knots and cut straight from the trunk. Even the best shaped trees could produce only half a dozen bows. With a standing force of at least 5,000 professional archers, and with the rest of the male population required to own and practise with a bow, the trees were needed in their thousands; perhaps their tens of thousands. It would have been far beyond the combined capacity of every churchyard in the country to supply them. It was proving, in fact, to be beyond the capacity of the country as a whole and from at least 1294, England was importing yew wood. So desperate did successive monarchs become to secure their supply that they enacted trading regulations requiring the attachment of a compulsory quota of yew staves to the importation of other goods. In a convenient way of passing the problem from monarch to merchant, Edward IV's 'Statute of Westminster' insisted that every ton of merchandise entering the country be accompanied by four bowstaves, a provision that Richard III expanded by demanding that every 'butt of malmsey' imported from Venice should be accompanied by ten.

Sheer quantity is a major argument against the use of church-yard yew for the provision of bows; but so is the matter of quality. Coming from higher altitudes or drier climates, continental trees grow more slowly, giving them a denser wood that is less brittle and less prone to knotting. It was always the continental tree that was the first choice of the English bowyer. An Act of Elizabeth I, in setting out maximum prices, demonstrates their comparative value: 'bows meet for men's shooting, being of outlandish yews, of the best sort not over 6s 8d; . . . bows being English yews 2s.' It was from foreign wood that the English were armed, and it was against countries that had supplied it that these arms would often be turned. Such is still the indiscriminate nature of the international arms trade. Yew wood came to England from across its entire continental range; from the Iberian peninsula to the eastern Carpathian mountains. Culled from lowland forest and mountain slope alike, it poured into the country from ports in Germany, Spain, France, Italy and the Netherlands. Reaching massive proportions, the trade produced one of the early ecological catastrophes with a whole species stripped out across the continent. Today, the churchyards of England and Wales are one of the last remaining reserves of ancient yews anywhere in Europe.

Given these arguments, modern commentators have come to be rather dismissive about stories connecting archery with the churchyard yew. They are, said Jennifer Chandler, in an article in *Folklore* in 1992, just 'an old man's fancy', her more radical reinterpretation of an 'old wife's tale', and 'pieces of antiquarian speculation from the seventeenth and eighteenth centuries which have been accepted into "lore".' And yet the stories persist, living on in tree talks and tree trails, in church guidebooks and in popular blogs. What Geoffrey Grigson calls the 'tallest tales about archery' have become part of the oral tradition. Whether old men's fancies, old wives' tales or elderly persons' imaginings, it might however be more revealing to dissect the detail than to dismiss them completely. It never hurts to have at least a little respect for the lore.

There is, first of all, the matter of those churchwardens' accounts. Writing in 1913, J.C. Cox put together a compilation of them covering some 300 years up to the late seventeenth century. At least some of them show entrepreneurial wardens obtaining an extra income for the church from the sale of yew wood. It forms, of course, just a footnote in the massive trade of the time; cheap wood sold, perhaps, for children's bows or for practice purposes, but it shows, at least, that the stories were not entirely fantastical. But there is, it seems to me, something even more revealing in them, and that is the way in which this single tree species had come, for a time at least, to represent a perception of the national character; to become, almost, the 'English' tree.

Practice with the bow was part of everyday life and all over the country boys were brought up on it. As Bishop Latimer put it in 1549, in a sermon before Edward VI: 'my poore father, was as diligent to teach me to shote as to learne any other thynge, and so I thynke other men dyd theyr children . . . for men shall neuer shot well, except they be brought up in it'. There was even an English way, he suggested, to handle the English bow: 'He taught me how to drawe, how to laye my bodye in my bowe, and not to drawe with strength of armes as other nations do, but with strength of the bodye.' It was the archers produced under such a regime that made England a military power, and a major player in the maelstrom of European politics. 'None but an Englishman could bend that powerful weapon the yew bow', as William Ablett expressed it in 1880. Notwithstanding the significant role of Welsh archers, or the foreign origin of the wood, the longbow archer was becoming a particularly 'English' phenomenon and his virtues to be seen as the virtues of the nation as a whole: strong, bold and resolute, disciplined and doggedly determined against the odds. It is a trope that recurs repeatedly in the national story, even into the twentieth century and the Second World War. In this creation of a national identity, the people become synonymous with the archer, the archer with the bow, and the bow with the wood that produced it. As the popular proverb put it:

> *England would be but a fling*
> *But for the yew and the grey goose wing*

The tree itself comes to stand for the national attributes, an idea picked up repeatedly by English poets, among them John Dryden who wrote, in the late seventeenth century, of,

> *The warlike yew, by which more than the lance*
> *The strong-arm'd English spirits conquered France.*

Later, with the decline of the archer and the rise of the navy as the basis of English military power, it was the oak that would supplant it as the 'national' tree, but for now it was the yew that was central in this mythologising of Englishness. Even that rebel archer, and man of the people, Robin Hood, was said both to have married and to have been buried under a yew tree. And here, too, was the origin of that populist passage in the pages of *The Gentleman's Magazine*. Such a romanticised role for the tree can only have been enhanced by its churchyard setting. Here, central to the life of every village, its steepled outline dominating both town and country, was this piece of distinctively English architecture with at its gate, an ancient, spreading and enduring yew. From the time of Henry VIII, of course, the church was to become, very literally, the 'Church of England', standing defiantly, like the archer, against the Catholic continent and the anti-episcopal Scot.

* * *

If the story of the yew and the longbow had not provided an explanation for the churchyard abundance of the tree, it had, at least, solved the other half of the mystery by providing the reason for its absence elsewhere. It had also given me another lead to follow. Writing in the first century AD, the Roman polymath Pliny produced what was probably the world's first encyclopaedia. In it he suggested that the Gauls went into battle with arrows that had been dipped in a poisonous decoction of yew. In Greek mythology, the goddess Artemis had used the same practice in hunting her enemies, while, according to some accounts, the early Irish used both yew and hellebore in a toxic mix on their arrows. The tree, then, is darkened by a double association with death; it was the means of mass production of a deadly weapon, and it is poisonous. This is the idea that Shakespeare expresses, with typical compaction, in the lines he gives to Sir Stephen Scroop in *Richard II*:

> *Thy very beadsmen learn to bend their bows,*
> *Of double-fatal yew against thy State.*

It is even possible that these two aspects have an etymological connection. The Scythian word for bow was *toxon*. It gives us *toxophily* for a love of archery, a word first coined by Roger Ascham in the sixteenth century. It could also have given us *toxic* and *toxin* and even, by some accounts, *taxus*, the Latin, and scientific, name for the yew.

The poisonous nature of the tree also gives rise to another of the theories set out in *The Gentleman's Magazine*. Its 'gloomy aspect' and its 'noxious quality', argues 'JO', serve as deterrents to any intrusion upon sanctified ground. The churchyard, in other words, was not there to protect the yew, the yew was there to protect the churchyard; its poison a deterrent to farmers or drovers who might be tempted to graze their livestock there. It is a theory favoured by Jennifer Chandler some two hundred years later and it appears again in the official government publication, *British Poisonous Plants*.

I have a copy of that document in front of me as I write. It is Bulletin No.161 of the Ministry of Agriculture, Fisheries and Food, published by Her Majesty's Stationery Office in 1968 at the price of 11s 0d. I came across it, I seem to recall, when the Islington Youth Service decided to dispose of its library and I have often innocently wondered for what purpose they were holding it in the first place. Its now rather stained and battered green cover is incised with the white outlines of some of the offending flora but the part that appeals to me most is the index. Here is listed a softly sinister litany of plant names, names that summon myth and mayhem, dark deeds in shadowy places and fairy tales from the forest: baneberry, belladonna, bittersweet, cowbane, hemlock, henbane, monkshood, nightshade . . . and so on, all the way through to wolf's bane. And to yew. 'All parts of the tree', the booklet tells me, 'are poisonous to man and to all classes of livestock' and 'the commonest symptom of poisoning is sudden death'. It seems rather inadequate as a diagnostic tool.

I sometimes suspect that Agatha Christie had read Bulletin No.161 before she wrote *A Pocketful of Rye*. It is one of her Miss Marple novels and in it we get a vivid, though not necessarily scientific, description of the effects of yew poisoning. The thoroughly unlikable Mr Fortescue, who happens to live in Yew Tree Lodge, is at his office one day when his secretary hears 'a strangled agonised cry' emerging from behind his office door. At the same time the buzzer on her desk emits a long and frenzied summons. Uncertainly she knocks on his door and enters to find her employer contorted in agony behind his desk, his body in convulsive spasm, his final words emerging only in jerky gasps.

'Tea – what the hell – you put in the tea . . .'

What is bad news for Mr Fortescue clearly delights the pathologist dispatched from St Jude's hospital. 'Very interesting case,' he tells Inspector Neele. 'Glad I was able to be in on it.' It was no accidental death, he asserts, but a clear case of taxine poisoning. It was berries or leaves from a yew tree that were added to the unfortunate Mr Fortescue's tea. He has met it before when some children playing doll's tea-parties stripped berries off a local tree to make their tea, but this is the first time he has encountered it as a deliberate act of murder. 'Delightfully unusual . . . You've no idea, Neele, how tired one gets of the inevitable weed-killer. Taxine is a real treat.'

Treat or not, Mr Fortescue's may not have been the first literary murder involving the yew. Some academics suggest that the mysterious 'hebenon' poured fatally into the ear of Hamlet's father was also derived from the tree. But I was grateful that it had given me the opportunity to read an entire detective novel under the only partially spurious guise of research. And Dr Bernsdorff was quite right about the taxine.

There are at least eleven of these alkaloid compounds in the yew, of which the most detrimental to human and animal health is the innocuously named taxine B. Apart from the 'sudden death' symptom, which is perhaps more applicable to livestock, the signs of yew poisoning include gastrointestinal problems, with diarrhoea and vomiting, a dangerous drop in pulse rate, possible heart failure and coma. The estimable Dr Lowe, who was, among other things, honorary physician to the future Edward VII, was part of a long and daring tradition of scientists who have tested such substances on themselves. 'I have undertaken a large series of experiments with taxin, made on myself at various times,' he comments mildly in his book. 'The tracings of the pulse show beyond doubt that it is a cardiac tonic of no mean value. The heart's action is decreased in frequency by small doses . . . at the same time that the cardiac pressure is distinctly increased. These effects I have found to be durable'.

It is perhaps a good thing for Dr Lowe that cases of human death from yew poisoning are actually rather rare. A 1992 article in the journal *Forensic Science International* quotes a global average death rate of one case every 3.3 years, with all of them to date having taken the form of suicide. I am tempted to wonder whether this includes the case of Catuvulcus, in 53 BC. According to Julius Caesar in his *Commentary on the Gallic Wars*, this King of the Eburones, in what is now Belgium, facing defeat, 'worn out by age . . . unable to endure the fatigue of either war or flight . . . destroyed himself with the juice of the yew tree'.

One assumes he must have extracted it from the leaves, for this is the most poisonous part of the tree, where over a third of the taxine content is comprised of taxine B. The situation with regard to the wood is less clear-cut. The heartwood of the yew is a beautiful orange-brown and I have seen the cross-section of cut boughs which take on an almost strawberry colouration. It contrasts beautifully with the creamy sapwood, the two together comprising a wood that is highly prized by cabinetmakers and turners. Among the many things for which it is recommended you will find 'cups, mugs and bowls', a suggestion which should come with something of a health warning. Writing in 1849 the Rev. C.A. Johns suggested that vessels of yew wood 'made poisonous the wine within them'. Perhaps he was echoing Pliny who noted in his *Natural History* that 'even wine flasks for travellers made of its wood in Gaul are known to have caused death'. Given this uncertainty, scientists at Kew set out to test the toxins in yew wood using a technique known as chromatography mass-spectrometry. These tests, which can detect trace amounts of specific compounds in even the most complex substances, failed to find any taxine B, but they did reveal the presence of an array of other taxines. The conclusions, set out in an article on the Kew website, are understandably cautious:

The toxicity of the taxine alkaloids detected in yew heartwood is not known. However, as the Kew chemists readily detected taxine contamination in wine into which yew wood had been placed, it would seem sensible to caution against testing Pliny's observation and not drink wine from a yew wood utensil.

There remains just one part of the tree that is not actually poisonous. As one of the earliest species of tree still around on the planet, the yew is unlike any of its coniferous relatives. The most obvious difference is that it does not produce cones. Instead, the female tree bears a bright, little, berry-like, flame-red fruit. Technically it is known as an aril for it is cup-shaped and open at the top to reveal a single black seed within. The seed is poisonous; the fruit is not. Given the tree's dire reputation, I have enjoyed the shock value of picking and eating the arils when out walking with friends, always remembering, of course, to spit out the pips. It is, I suppose, a rather petty party trick and I have been placed under a strict injunction not to perform it in front of children, though I don't, personally, doubt their ability to master the appropriate expectoration. The berry has a rather mucous quality, with a sweet and slightly fragrant taste. The presumably fatal children's tea party, from which Dr Bernsdorff learnt so much, must have involved a very considerable number of arils in the pot, complete with all their pips. Fred Hageneder, who has written the very best of all the many books on the yew, quotes a German source in saying that to be seriously affected a child would need to eat between sixty-five and one hundred of them. They would also need to chew them thoroughly, for swallowed whole they would probably pass straight though the system, which is, after all, how a seed within a sweet container is usually disseminated. But this information, too, I will probably keep from the children.

I was discussing my findings to date with my sister when she related a curious tale. She had been, for many years, a social worker in Havant in south Hampshire and one day, it happened to be her birthday she recalled, she had received a call from a children's home concerning a young man who was in her charge. He had, it seemed, attempted to poison the entire occupancy of the place by picking yew berries, stealing into the kitchen and putting them in the shepherd's pie. Perhaps the potential enormity of the offence had come home to him, for he had confessed it to the staff before the dinner was served. He was appropriately contrite by the time my sister got there and, she was at pains to tell me, he grew up into a 'fine young man'. A youth worker, in fact.

The next time I went to visit my sister, and still being under the influence of Ms Christie, I persuaded her that we should visit the scene of the putative crime, which was set on Hayling Island. There is only one road onto the Island and it is always congested. Indeed, with a plan to build 700 new homes there it may soon become completely inaccessible. The children's home, at the southern tip of the island, had long since been demolished, but the site was still there, surrounded by a high grey hoarding. Above it, I could see the tops of the trees that had once graced its grounds: a tall pine, a large spreading cedar, and beside what would have been the entrance gate, a disappointingly small yew. What's more, it was an Irish yew. It's not that I've anything against them but the Irish yew is a cultivated form of the tree, growing in tight, compact and shrubby shape, with upright little spires that give it the resemblance of a miniature Orthodox cathedral. It is the yew now most commonly planted in cemeteries and is often marketed as the 'churchyard yew'. All Irish yews originate from just one specimen found growing wild in County Fermanagh in 1780 and, bringing all my forensic abilities to the fore, I recalled that they are always female. This, then, could well have been the source of those dinner-bound berries. But it wasn't actually bearing fruit and not far away was another candidate, for, set in the nearby churchyard of St Mary the Virgin, the parish church of South Hayling, was one of the country's famous ancient yews. Wanting to cover all bases, we went there too.

The church was large, and locked. Built of multi-coloured stone, white, red or honeyed, it was pierced with lancet windows and topped with a hipped and slate-covered spire. In its grounds the long yellow grass of late summer clustered round the tombstones, except in some further reaches where close-cut lawns awaited their allotted portion of the dead. Seeking out the specific tree we bumped into a couple set on the same mission, having read about it in a leaflet on 'local attractions'. My sister, as is her way, engaged them in conversation, explaining our own visit and attempting to sell a copy of a book I had not yet written.

Literary descriptions of specific yews tend to wrestle with superlatives. This was, indeed, a magnificent tree, but there is no getting round the fact that the ancient yew is also extremely untidy. It sprawls. It falls. It leans over in every direction, sprouts from every surface, rots here and there, splits into peeling, twisted and fluted trunks, and forms indecipherable tangles of twig, bough and branch. So it was with the South Hayling tree. It was like an elderly dowager, supreme in her self-confidence and no longer having to put on an appearance for anyone. She seemed, too, to walk on sticks, being propped up with poles at various ungainly angles. She had, at one time, been surrounded with iron railings, at the cost of a parishioner who now lay buried beside her. With supreme contempt she had leant down in various places to bend them out of shape and was threatening to crush them completely.

There was, in her shade, a memorial bench where I sat to make my notes, rather resenting the overflowing concrete bin beside me which seemed an insult to the aristocratic lady. The bench was dedicated to the interestingly named 'Moss Dollery 1926–1985' and the inscription was in French: *A Notre Ami Que Nous N'Oublirons Jamais. French Parties 1983–84–85.*' I did not know what the 'French Parties' might have been but I wished I'd been invited. Though the tree was indeed female it was bearing no berries. The church grounds, however, contained no less than twenty-five other yews, of all shapes and sizes, and some of them with the fiery red fruit. I persuaded my sister to try some and we stood, sucked and spat companionably for a while. I refrained from telling her some of their country names of snotty gogs, snotter galls and snottle berries. It wouldn't, I thought, have increased her enjoyment. Perhaps they were, I wondered, from the very same tree as those which had once polluted a children's centre shepherd's pie.

Notwithstanding these uses, both fictional and factual, I was confident that the poisonous quality of the yew was not the reason for its churchyard setting. There is, to begin with, a debate about its actual effect. Whilst yew can be fatal to cattle, sheep and horses, those that graze on it regularly, or eat it in smaller quantities, seem to become inured to it. In some places it was even added as a supplement to winter feed. More significantly, planting such a slow-growing tree would hardly be particularly potent as a protective device, especially when most churchyards and cemeteries were already walled or fenced. Nor would it be conducive to good relations with the human flock for the parson to take such aggressive action against their animal charges. Many, in fact, seem to have adopted the opposite approach, allowing grazing in the churchyard, perhaps encouraging it as a way of controlling unruly plant growth. Some even obtained an income from it, as an article in *Nature* in 1944 points out, issuing licences to feed cattle in the church grounds. Long before it became the separated precinct of Victorian piety, the medieval churchyard was a far livelier place, and more integrated into the secular life of the village; a setting for meetings, fairs and markets. The idea that yews were planted there to protect the place from incursion is the product of more modern sensibilities; it is not the explanation for the churchyard yew.

* * *

In my investigation so far I had considered myself to be standing on relatively solid ground. But the world of yew research is not like that. It consists of confusions and contradictions, assumptions and assertions, differences and disagreements. It is rife with speculations on a distant spirituality and imaginative reconstructions of a supposed pre-history. It becomes, in short, an exercise in uncertainty and dislocation. Perhaps this is appropriate for a tree whose 'noxious influence' seems to exert itself over the mind as much as the body. In 1970 a medical professor, recently retired from the University of Greiz in eastern Germany, started to hallucinate whilst gardening under some yew trees. After a period of dizziness and nausea he began to experience scenes of vampires and vipers. They were followed by visions of paradise and a final euphoric happiness. His later, more sober, conclusion was that on warm, still days the tree emits its toxins in gaseous form, lingering in the shade with a hallucinatory influence. Could this, I wondered, be an explanation for the thread of dreams and visions that recur throughout the literature and folklore of the yew? They can be innocent enough, like the suggestion in the *Dictionary of Superstitions* that a girl who picks a sprig of yew from a churchyard she has never before visited, will dream of a future lover. Sometimes they are spurs to the creative imagination, as in the disputed account that Mendelsohn, still only 17 years old, composed the music for *A Midsummer Night's Dream* sitting under a yew tree at night. But they can be darker and more disturbing. The poet and prophet, Thomas the Rhymer, met the Faery Queen through an hallucination brought on by eating yew berries, while in *The White Goddess* Robert Graves gives a depiction of the 'Night Mare', an aspect, he says, of the goddess, which builds its nest in the branches of an enormous hollow yew, lining it with white horse-hair, the plumage of prophetic birds and the jaw-bones and entrails of poets.

Shakespeare seemed to understand this, as well as every other aspect of the yew, a tree he mentions in at least five of his plays.

In *Romeo and Juliet* the ill-fated hero returns illicitly from exile, entering the graveyard at night to find the tomb of his supposedly dead lover. He dismisses his servant Balthasar with an injunction to leave the site but, concerned for his master's safety, Balthasar disobeys him. He sits down under a yew and dreams, until disturbed by the arrival of Friar Lawrence, who has come to awaken Juliet from her drugged state. He tells the Friar:

> As I did sleep under this yew tree here
> I dreamt my master and another fought,
> And that my master slew him.

It is a report that turns out, tragically, to be true.

Within the context of these strange and airy influences, it is curious that one of the best-known yew researchers of recent times, and the main proponent of a pre-Christian and Druidic connection for the tree, was driven in his work by a sequence of dreams and visions. A remarkable tree calls forth remarkable characters and Allen Meredith is one of them. Having left school at fifteen without qualifications, he joined the Royal Green Jackets and served under the United Nations in Cyprus. Later he had try-outs as a professional footballer, spent several months living in a wood and walked the roads of Oxfordshire promoting tree-planting projects with local schools. At some stage during all this the dreams began, the first involving a group of cloaked and hooded figures with hidden faces who instructed him to seek out the true 'Tree of the Cross'. In another he was 'given a word', a word which he never reveals, but which is 'the whole reason why I'm doing this work'. It was work that included cycling round the country finding, measuring and recording hundreds of yews. As he amassed more and more information, and the intermittent dreams continued, he came to ascribe to the yew an immense and almost universal spiritual significance. It was one, he believed, on which the future of humanity depended.

'The yew tree is the most sacred thing on this earth, here for some very special reason. If we don't recognise [this] I don't think we will survive . . . The yew tree is part and parcel of each one of us. It is obviously a divinity.'

His ideas were expounded in *The Sacred Yew*, a book written by Anand Chetan and Diana Brueton, two of his followers. The historians have it wrong, they explain, in seeing the oak as the sacred tree of the Druids when, in reality, it was the yew, and the many myths ascribed to the apple also rightly belong to the red berry of the yew. From here on, the claims come quick and fast. The yew cult was the first human expression of religious awe and the great circular monuments like Stonehenge and Avebury are based on the natural shape of the yew grove. The Battle of Hastings was fought under a yew tree, the Magna Carta was signed, not at Runnymede but under the yew tree at Ankerwycke and the early English kings were inaugurated beneath its spreading boughs. The yew was the origin of the yule log and the source of Frazer's *Golden Bough*. It gave rise to the phoenix legend. It is at the root of the word 'Jehovah'. It is the biblical 'Tree of Life' and the wood of the true cross. It is Yggdrasil, the World Tree, the Axis Mundi. And it is a repository of ancient wisdom brought to Britain by the 'wise people' for safekeeping, giving this country 'the greatest concentration of sacred trees in the world'.

As for the churchyard yew, the connection becomes clear. It is not the yew that came to the churchyard but the churchyard that came to the yew, as the new Christianity colonised the ancient faith and built its churches in its sacred groves. Meredith made extravagant claims, but he had, at the same time, almost single-handedly brought the yew tree back to public attention. He had produced a gazetteer of hundreds of ancient yews, all of them painstakingly measured; investigated archives and uncovered new historical material; campaigned for the tree's preservation and railed against its lack of statutory protection. He had also exerted a considerable influence, with some eminent botanists and arboriculturalists, expounding his thoughts on an international stage. David Bellamy appears to have fallen particularly under his spell, providing an introduction for *The Sacred Yew*, and seemed particularly convinced by his arguments on the dating of yews. And so, too, was the Conservation Foundation, which Bellamy had founded, issuing hundreds of certificates to attest the great age of yews around the country. They include the one at South Hayling, rather faded now and attached to the railings, next to another which declares that all accoutrements to graves are illicit and must be removed forthwith. The certificate declares the tree to be 2,000 years old and its four signatories include both Allen Meredith and David Bellamy.

Meredith was not the first, however, to suggest a pre-Christian origin for the churchyard yew. Once again, *The Gentleman's Magazine* was early in the field. 'We may fairly conclude', wrote 'TO' in 1781, 'that the custom of planting the yew in churchyards took its rise from Pagan superstition, and that it is as old as the conquest of Britain by Julius Caesar'. The Victorians enjoyed something of a romance with Druidism, adding much embroidery to their vague history, and this was especially reflected amongst the poets. The opening to Martin Tupper's 1855 poem on 'The Service-Yew on Merrow Down' is a typical example:

> When the Druid, long of old,
> Solemn stalk'd in white and gold
> Down among those ancient yews
> Ranged in serpent avenues,
> Then wert thou a sapling tree,
> Then that Druid planted thee,
> Thousand-winter'd son of earth
> Thirty feet around in girth!

After having read several similar examples I found it hard not to wish that a few more of those poets' entrails might be found littering the Night Mare's nest.

But the idea had taken hold and my much-consulted copy of Mrs Grieve's 1931 *A Modern Herbal* shows that you can't keep a good story down, whatever its veracity. 'The yew', she says, 'was a sacred tree much favoured by the Druids who built their temples near the trees – a custom followed by the early Christians.' The connection became a sort of commonplace, repeated in volume after volume. As it still is today. Tony Hall's *The Immortal Yew*, published in 2019 by no less an authority than the Royal Botanic Gardens, tells us that yews were 'sacred to the Druids who erected temples close to them' and that 'many of the churchyards where ancient yews grew were originally sites of Celtic or Druidic worship'. It was not the first time that Kew has wandered in this direction. Roy Vickery in his *Folk Flora* quotes a label that appeared on a yew tree in the Botanic Gardens in 1993. 'The Druids', it read, 'regarded yew as sacred and planted it close to their temples. As early Christians often built their churches on these consecrated sites, the association of yew trees with churchyards was perpetuated.' Thomas Pakenham, in his otherwise excellent *Meetings with Remarkable Trees*, gives a particularly lurid account. 'Once its branches', he says of the tree in Much Marcle churchyard, Herefordshire, 'might have carried pagan trophies, or the severed heads of sacrificial victims. Christianity would have purged it of that'.

Back in the 1890s the phlegmatic John Lowe was having none of it. 'Several writers', he reports, 'have entered into this question, and presume that the yew was one of those evergreens which, from its shade and shelter, was especially cultivated by the Druids in their sacred groves and around their sacrificial circles . . . and that in this manner arose the association of the yew-tree with our churches and church-yards.' The main problem with the theory, he says, is that he hasn't found a single piece of evidence to support it. And there, bluntly expressed, is the main objection to the theory of a Druidical connection. Of those much asserted 'temples', not a single remnant has been found; no sacrificial altars and no cultic implements. In fact, as Ronald Hutton argues, 'not a single artefact has ever been turned up anywhere which experts universally and unequivocally agree to be Druidic.' Though there is plenty of material evidence for people with religious knowledge and skills in Iron Age Britain, there is little for a specialised priesthood. 'We may need to scrap the Druids from Iron Age archaeology,' he concludes. 'They really seem to be doing more harm than good.'

What little information we do have about the Druids comes from written sources, from the biased accounts of invading Romans and from Ireland. They paint very different and contradictory pictures. There is not even, anywhere, an agreed definition of the Druid. As Roy Vickery points outs 'Current scholarship suggests so little is known about the Druids that most of what has been written about them simply reflects what the writers would like them to have been, rather than what they actually were. Much of what has been written about yew is probably best forgotten or ignored.' A very large castle, it seems, has been built on a very flimsy foundation.

Apart from lack of evidence, there is another problem with the Druidic theory and it relates to a vexed question that researchers have been arguing over for generations: how to determine the age of a yew. No other tree puts so many obstacles in the way of its accurate dating; and no other tree has seen so much effort expended in trying to resolve them. You cannot count the rings of an ancient yew, for hollowing out is a part of its natural growth process. You cannot use carbon dating for the similar reason that none of the oldest wood is available. You can try to measure the girth and consider growth rates but this too is fraught with difficulty. Old trees fork low or break into multiple stems, they have burrs, bumps, outgrowths and hollows. They may even be fragmentary with the surviving portions of growth now separated. The simple question of what, where or how you measure it becomes a minefield. And should you arrive at an appropriate approach then your further calculations will be complicated by the fact that growth rates vary according to multiple environmental factors, including waterlogged soil, severe winters, late frosts, cold winds, exposed conditions or drought. Growth can slow down, restart, and then slow down again and some old yews seem to have given up on growing altogether, or, at least, until some significant change in conditions sets them off again. Thus the Totteridge yew in north London has, despite repeated measurement, shown no growth at all in its girth since 1677 – a period of more than 350 years. It is not surprising, therefore, that the age estimates of our oldest yews have fluctuated wildly, sometimes by a matter of millennia.

Across the last 200 years the aging of our yew trees has operated like something of a chronological yo-yo. Victorian botanists began to ascribe very early dates to yews, a process encouraged by church guides and local history books all eager to proclaim that their yew had been there when 'Julius Caesar arrived on these shores' or 'when Jesus walked the earth'. Twentieth-century researchers developed new approaches that lowered these more extravagant estimates, and this remained the norm until the work of Allen Meredith began to raise them all again. Though less scientific, his dating techniques, based, he said, on intuition, experience and a great deal of circumstantial evidence, proved influential and once again our yews became much older. It was on these estimates that the Conservation Foundation certificates were based, but by the beginning of the new century these were already coming under challenge. Robert Bevan-Jones was among the first to argue for much later dates, suggesting that our oldest yews went back no further than the Saxon period, and about the same time the Ancient Yew Group was coming to a similar conclusion.

Probably our pre-eminent figures in yew research, they had formulated a Unified Field Theory which tried to eliminate conflicting results through a complex formula combining different approaches, including tree girth, tree rings and historic rates of growth. It led to some radical reassessments. The Aberglasney yew tunnel in Carmarthenshire, once thought to be 1,000 years old, proved to be a mere stripling of 250. The Llangernyw yew in Conwy came down from 5,000 to 1,600 years; the St Cynog's yew in Powys from 4,000 to 1,200 and the famous Ankerwycke yew from 2,500 to a comparatively sprightly 900, thus ruling out those pre-Norman coronations. These are still great ages, of course, and these trees should remain an object of our awe; but they are not a lineage that takes us back to Neolithic times. And those Conservation Foundation certificates probably need some redrafting.

* * *

The Fortingall yew in Perthshire is one of the few trees that still retains an age estimate of more than 2,000 years. Perhaps the tree with the best claim to be the oldest in Britain, and possibly in Europe, it is surprising that it should be located in Scotland, where ancient yews are otherwise very rare. But its position there is propitious. Set at what many consider the very centre of the country, it sits at the Head of Appin, a point where three glens meet, beneath the great bulk of Schiehallion. It is an area surrounded by Bronze Age and Celtic relics.

This was a tree that I very much wanted to see and I set out to find it one grey and windy morning with two friends from Edinburgh. Ann and Stuart, who would be my guides, also happen to be members of the local branch of the Fortean Society, the organisation that investigates all things abnormal, paranormal or downright odd. The day worsened as we drove north-west, with a steady chill rain, and though it was suitably atmospheric, one might have wished for something finer from

July. From Aberfeldy, with whose whisky I am well acquainted, we followed the River Tay along the Appin of Dull, a valley of barley and potato fields and of quietly grazing sheep, with gentle but extensively forested hills rising on either side. Halfway along was the village of Dull, whose sign proclaims it: 'Dull. Twinned with Boring, Oregon'. It was, of course, too good a photo opportunity to miss. The village of Fortingall itself sits beneath Balnacraig and the lower reaches of Schiehallion. It is small and neat with the sort of stylistic unity that suggested it must have been the work of some beneficent laird. As indeed it was. Donald Currie, having made his money in shipping, had bought the estate in 1885 and rebuilt the village in a sort of proto-Arts and Crafts style. Even the pub, and the church adjacent to it, matched each other in their whitewashed walls and crow-stepped gables. The castellated wall of the churchyard is dressed with white lichen and capped with dark green moss, and balanced on its gateposts are two curious water-worn stones, something of a common feature in the area. Though the church is a relatively recent construction, the tomb-stones within its grounds are older, weather-beaten and lichen-encrusted with one table tomb so thickly topped with moss that I fought an urge to lie down and stretch out on it. Leaning up against the church wall were stones inscribed with crosses, and an old font removed from the earlier church building. Somewhere in these grounds, in an unmarked grave, lies the body of the Campbell colonel who had ordered the inhospitable massacre of Clan MacDonald at Glen Coe. One hopes he sleeps uneasily. A grey slate path with incised inscriptions leads through the churchyard to the tree, indicating that it has long been an attrac-tion for visitors, and an interpretation board is there to tell them that it is somewhere between 2,000 and 9,000 years old. Perhaps a rather large margin of error is in order given the constantly changing currents of yew tree dating. Despite its age, for even the lower estimate is impressive, it is quite likely that a good portion of Fortingall's visitors will go home disappointed; for what greets them is no magnificent spectacle, no striking silhouette, no huge sweeping boughs and no huge bole of almost unimaginable girth.

Once, of course, it was very large, and in 1769 Thomas Pennant, one of Gilbert White's regular correspondents, measured it as fifty-six and a half feet, or just over seventeen metres, in circumference. So much of it has now rotted, however, that it survives only as two fragments, set so far apart that they seem like separate trees. In a further reduction to the spectacle, they stand half-hidden behind a high wall, pierced with barred openings through which you must gaze as if talking to a prisoner in his cell. The wall first went up in 1785 and was rebuilt in 1842; its purpose to protect the tree from the depredations of visitors who were cutting branches to take home as souvenirs or good-luck charms. But it was not only the visitors who were to blame; there also seems to have been a minor local industry thriving from it. Visiting in 1883, a Dr Neill described how it had suffered 'considerable spoilations' including 'masses of the trunk itself carried off by the country people, with the view of making quechs, or drinking cups and other relics which visitors were in the habit of purchasing'. One can only hope that they didn't fill them with wine.

Despite these indignities, the remaining portions of the tree seem still to be vigorous, with fresh green growth springing from their otherwise silvery darkness. Some branches loop over and lean on top of the wall, whilst one substantial bough is supported by a stone pillar. There are the usual untidy twiggy tangles and, beneath them, a dense shade where only ground elder will grow, amidst which a robin was feeding its wing-shivering fledgling. It is a peculiarity of the tree that although it is a male it has, in recent years, been producing a single branch that is female. Sexuality is more fluid in the yew, and in many other trees, than most of us realise. Justifying the latitude in the signage, the

Unified Field Theory had set the age of the tree at around 2,000 years old, lopping 3,000 years off the estimate by Allen Meredith. Visually spectacular or not, it remains a remarkable organism. Trees of this age, says the Ancient Yew Group, are 'rare [and] staggeringly important in national or even international terms.' It is a tree that genuinely can live up to those claims about Jesus and Julius Caesar.

* * *

Standing in the shadow of Schiehallion, literally the 'fairy hill of the Caledonians', the Fortingall yew is surrounded by a range of other ritual and sacred sites, many of them pre-Christian in origin. Within the churchyard itself cup-marked stones have been dug out from between the roots of trees while the surrounding fields contain at least three stone circles. Across the road from the church is a Bronze age burial mound, atop which a fire was lit every year at the Celtic festival of Samhain, a practice that continued as a community event until 1924. Other sites, which Ann and Stuart were keen to explore, lay along the great length of Glen Lyon. Clearly, the whole area had once been of considerable social, spiritual and political significance and set at its geographic heart, and at the very head of Scotland's longest glen, the location of the Fortingall yew can hardly be coincidental. The suggestion of such ancient and potentially 'pagan' associations of a churchyard yew would prove anathema, however, to the more conservative clergy within the established church and, to escape such a taint, Christianity was providing its own explanations for the churchyard locus of the yew. Some were elaborately theological, as with the suggestion that the red heartwood of the tree represented the blood of Christ, and the white sapwood his body, thus neatly mimicking the elements of the mass. The more common explanation was a more practical one. It concerned the use of the yew in celebrating a part of the Easter story.

The idea of marking the Sunday at the beginning of Holy Week with a palm-bearing procession goes back as far as fourth-century Palestine. It was a recreation of the triumphant entry of Jesus into Jerusalem, riding on a donkey but hailed by the crowds as a king. According to John's gospel they greet him by waving branches of palms and the day officially became 'Palm Sunday'. By the seventh century the practice of processions had reached Britain, where it continued to grow in popularity despite being banned under two monarchs.

The Palm Sunday processions are still with us, practised even here in our urban East End parish. The servers and the clergy take the lead, resplendent in their robes of white and red and bearing a large wooden cross. They are followed by a straggling hymn-singing congregation as they make their way through the streets, estates and underpasses. It is the common practice for the people to carry large fronds of palm, as well as small plaited-palm crosses, the problem, before such things could be imported, being what to use in its place. The goat willow was one substitute, its golden catkins bright at this otherwise stark time of year, another was the evergreen yew. As William Caxton wrote in 1480: 'Wherefore holy church this day maketh solemn procession in mind of the procession that Christ made this day – but for reason we have no olive that bears green leaf always, therefore we take ewe instead of palm ... and bear about in procession.'

So widespread did this usage become that in Ireland, Devon and parts of southern England the yew was commonly referred to as 'the palm', while in Ireland, Palm Sunday often became 'Yew Sunday'. It was only a short step from here to suggesting that the yews had been planted in the churchyards deliberately to meet this need. The idea occurs in the writings of the seventeenth-century diarist John Evelyn and is argued over in the pages of *The Gentleman's Magazine*. 'AB' puts it forward in 1779 but is rebuffed by T. Rowe who objects that 'the bearing of palmes on Palm-Sunday was an act of joy and ovation in remembrance of our Saviour's triumphant entry into Jerusalem; whereas the yew is not only a tree of baleful influence but is too much of a funereal nature to be made a substitute for the joyful palm'. AB, however, is having none of it and responds that the tree is symbolic of resurrection and, therefore, 'the funereal use does not make it inappropriate'. Besides, he argues, 'many funerals were merry occasions.'

It is not an idea, however, that stands up to much scrutiny. The custom of using yews in place of palms was restricted to particular parts of the country, yet the actual distribution of the churchyard yew is much wider. The oldest yews actually predate the period of the procession's greatest popularity – and if the yew could be planted for this purpose, then why not the goat willow which was more widespread in its Palm Sunday usage but almost never occurs in a churchyard? The yew was not planted for Palm Sunday; it was used because it was evergreen, because its cut branches could form regal fronds and because it already stood conveniently next to the church.

* * *

In the early months of 2020 I was making a series of programmes for Radio 4 with Brother Sam, a Franciscan friar and friend of mine. The programmes followed the events of the Easter story through the wild flowers and trees with which they had become associated in British folklore. Because of its Palm Sunday connections, the yew was one of them. Among the places we visited was St Peter's church, just outside the village of Tandridge, on the sloping shoulder of a gentle hill in greensand country at the edge of the Surrey Weald. We drove there on a grey day, down lanes lit by white-flowering hedgerows of blackthorn. Red kites were calling plaintively as we recorded in the chill churchyard and we completed our work huddled together with Julian our producer, inside the hollowed heart of the tree, marvelling at how its silence seemed to absorb every sound around us.

The Tandridge yew is another of the few still dated at over 2,000 years and it made such an impression on me that I took some of my family to see it in the summer, once the long period of Covid lockdown had sufficiently eased. The kites were still calling but the daffodils and primroses of spring had given way to ox-eye daisies, white clover and the red and yellow flowers of bird's foot trefoil, justifying, my wife remarked, its country name of 'eggs-and-bacon'. It was all drowsiness. A chiffchaff called monotonously, the sound of a dripping tap, a song thrush sang repetitively from a sycamore and wood pigeons repeated their endless melancholic lament. Impressive shuttlecocks of male fern reared up between the tombstones and dancing pairs of comma butterflies mated in mid-flight, one breaking off to bask on the table tomb of 'Colonel Gerald Leslie Pepys C.B., D.S.O. Indian Army. A Just Life Made Perfect'. We found a path to one side of the churchyard that led through a copse to reveal a glimpse of the one-time manor house. Whoever had been the occupants, this had clearly been their own private path to the church in the days when God was not a democrat.

The church itself was an attractive stone structure with a modest tapering spire clad in wooden shingles. The yew stands right next to it, at its eastern end, close enough for its spreading boughs to tap gently on the church roof. The Ancient Yew Group describe it as 'one of the most spectacular specimens in the world' and when they took a party of German yew enthusiasts to visit it once, they observed a five-minute silence. Both the praise and the respect are justified, for this is truly a remarkable tree. It rises from the ground like a continuous wall of wood, but so deeply fluted it could be a giant organ case. At its centre sits a little porthole that looks into the hollowed-out interior. It had once, for some reason, been concreted up, but blessedly, the concrete is now falling out again. A little further up, the tree divides into three main trunks, as well as several smaller ones, and the deep fluting continues up them becoming like muscles and sinews in an anatomical drawing. Earlier writers had speculated that it had originally consisted of several different trees that had somehow fused into a single growth, but this idea is now rejected. I confess to not normally being a fan of Wordsworth but his description of the yews at Borrowdale seemed apt:

> *Huge trunks! – and each particular trunk a growth*
> *Of intertwisted fibres serpentine*
> *Up-coiling and inveterately convoluted!*

I walked round the tree to reach the entrance to its hollow interior. It was charred from where bonfires had once misguidedly been lit, but the wood beside the opening was beautifully patterned, with flowing and winding curves as though it had been marbled. This was, I had learnt, the form taken by 'secondary wood' that grows over and strengthens the original layers as support for the increasingly thin-walled trunk. From this side of the tree you could see more clearly how its great branches leaned out and down around it, until they reached the ground and rooted themselves. What looked like young yews growing in a circle were, in fact, new growths of the old tree springing up as a whole copse around the original.

I had read on a local website that the tree had been 'immortalised' on a first-class stamp. Checking this out I found that in 1999 the Post Office had issued forty-eight new stamps to commemorate the Millennium. The next year it had issued forty-eight more to celebrate 'Millennium projects'. One of these shows the Tandridge yew. 'Immortalised' seemed an inappropriate word for the relationship between a collector's piece of postal ephemera and a 2,000-year-old tree. I was sharing this opinion with the tree when I happened to notice one final detail: growing out of it, some at the base and some on higher branches, were the bulbous brackets of an almost garish egg-yolk yellow fungus. It was, I was excited to realise, the sulphur polypore, more enticingly known as the 'chicken-of-the-woods'. Enlisting the help of the family, and asking permission from the tree, we cut some and bore it triumphantly home.

The next day happened to be Father's Day and my eldest son, who was staying with us at the time, cooked it as a breakfast treat. Eschewing all the fancy recipes I had in books, he cut it into chips and fried it up with eggs and vegetarian sausage. It was, perhaps, more interesting, certainly more chicken-like, in its texture than it was in its taste and the family was divided as to its merits, variously describing it as like microwaved eggs, polystyrene cutlets or soya chicken. Personally I enjoyed it, and the gift of being cooked for, and was close to finishing when it occurred to me to wonder whether mushrooms cut from the poisonous wood of the yew might not themselves have absorbed something of its toxic quality. I refrained from saying anything to the family but I spent an anxious few hours waiting for coma or convulsions to commence. It would, I thought, have been an unfortunate Father's Day if I had managed to eliminate the entire family.

* * *

From Murthly Castle, on the right bank of the River Tay near Dunkeld, an avenue of seventy yews leads to a chapel. According to tradition, the laird can walk the avenue from chapel to house as often as he wishes, but he can never walk it in the opposite direction. It is known as the 'Dead Walk' and the only time he passes that way will be in a coffin. Notwithstanding my own family's survival, it seemed to me that in the story of the yew, death had been the consistent and recurring theme. I was constantly coming across it in my visits, my researches and my reading. It ran like a thread through poetry, folklore, custom and history. It was even there in archaeology. The yews at Murthly are only two hundred years old, the tradition there is not ancient, but it has its antecedents in much earlier connections between the yew and the dead.

The British countryside is dotted with barrows: long barrows, round barrows and oval barrows, ball barrows, bell barrows and saucer barrows, the humps and tumps and tumuli that form the burial mounds of the Neolithic peoples. They appear on remote downland sites but also on town commons, in wide open country but also in closely farmed fields, where they are sometimes ploughed almost out of existence. Many of them, according to records, had once been planted with one or more yews, even though some have subsequently been felled, or even cleared for archaeological excavations. They include such places as Duck's Nest Barrow in Hampshire, Cranborne Chase in Wiltshire, Wormelow Tump and St Weonard's Tump in Herefordshire, and Taplow Court in Buckinghamshire where the hoard of gold and silver artefacts was second only to that found at Sutton Hoo. A barrow with a yew tree on it stands in the churchyard at Ashbrittle, Somerset, another once stood above the chambered tomb at Llanhamlach near Brecon, and at Denny Lodge, in the New Forest, seven barrows occupy the area known as 'Yew Tree Heath'. In 1930 the archaeologist A.S. Newall excavated a barrow at Amesbury in Wiltshire. Inside it he found a grave lined with the wood and leaves of yew. The body, he concluded, had been heaped on burial with yew branches and moss.

Similar practices appear in the later traditions of all three British nations. I had already encountered them at Fortingall where it had been the well-documented custom to carry coffins through the central hollow of the tree on their way to the churchyard. As the coffin was lowered into the grave, the mourners would hold branches of yew above it, then toss them in as the grave was filled. Such practices must have occurred in at least some parts of Wales, for Andrew Morton, in his book on Welsh yews, recalls as recently as the mid-1980s seeing a sprig of yew being placed into a freshly dug grave. 'It is customary', the gravedigger told him, 'and would wish a good passing to the deceased'.

The widespread nature of these practices is attested by the Rev. J. Collinson, writing in 1656 on his visit to the Ashill yew. 'Our forefathers', he tells us, 'were particularly careful to preserve this funereal tree, whose branches it was usual to carry in solemn procession to the grave and afterwards to deposit therein under the bodies of their departed friends'.

Vaughan Cornish, whose 1946 book on yews had been very influential, suggested that the location of yews in churchyards was based around funerary practice and that there were often two of them; one by the main gate with its coffin rest, the other by the path leading to a second doorway and following a route described in the Book of Common Prayer.

Such funerary connections were by no means restricted to Britain. They feature in the burial traditions of the Celtic and Germanic tribes and appear again among the Phrygian, the Saxon, the Merovingian and the Georgian peoples. In ancient Egypt, the yew was the foliage of mourning, as it was along with the cypress in Greece and Rome, where a branch of yew would be used to light the funeral pyre. For Ovid, it is yews that line the route travelled by the dead to Tartarus, the particularly unpleasant part of Hades. It is a suggestion echoed in an Austrian tradition, where on All Saints' Day, branches of yew were brought to tombs of the recently dead, to guide them on their return to the land of shadows. Further afield, the yew is to be found on the sacred sites of some of the native American peoples, where it is regarded as the gatekeeper, connected with birth as well as with death. In Japan it grows on graves as well as at Shinto and Buddhist shrines, and on the Alaskan peninsula the Tlingit people use yew wood to carve death masks and spirit whistles.

Here in Britain, the intermingling of death and the yew provides a rich vein running through the course of English poetry, and Shakespeare addresses those funerary rites directly.

In the second act of *Twelfth Night*, the Duke bids the Clown to sing for him, a song which seems almost incidental to the plot. But it is a beautiful, melancholic piece of poetry:

> *Come away, come away death,*
> *And in sad cypress let me be laid.*
> *Fie away, fie away breath,*
> *I am slain by a fair cruel maid:*
> *My shroud of white, stuck all with yew.*
> *O prepare it.*
> *My part of death no one so true*
> *Did share it.*

It is, says the Duke, a common enough piece, sung by 'the spinsters and the knitters in the sun' and the carefree maids that 'weave their thread with bones'. A poem from 1647 by Thomas Stanley, the man regarded as the last of the metaphysical poets, makes the same link between the cypress and the yew:

> *Yet strew upon my grave,*
> *Such offerings as you have*
> *Forsaken cypresse and sad Ewe*
> *For kinder flowers can take no birth*
> *Or grow upon such unhappy earth.*

Following on from Shakespeare and the metaphysicals, the theme appears in Wordsworth in his poem on the yew trees of Lorton Vale, and again in Tennyson whose 'Old Yew', 'graspest at the stones, that name the underlying dead'. It is there, too, in the work of the Scottish poet John Leyden and in A.E. Housman's *Shropshire Lad*. Walter de la Mare has the yew 'burning its lamps of peace for those that lie forlorn', while for Sylvia Plath it takes much darker form. 'The message of the yew', she says, 'is blackness – blackness and silence.'

In these words Sylvia Plath had given me the clue I was seeking. It is hard now to reconstruct, even imaginatively, the relationship people would have had, until comparatively recently, with plants and trees and animals. Their lives depended on an intimate acquaintance. They were their food, their friends, their foes, their shelter and their medicine. They were the backdrop to everything they did, the adornment to their daily lives and their compatriots in mortality. They would have been familiar with their characteristics, and known their characters, and used them, writ large, in their art, their storytelling and their spirituality. To understand our historic relationship with the tree, and how it gave rise to the churchyard yew, we must understand those characteristics. None of them is more immediately apparent than that darkness and silence; and those, too, are the apparent characteristic of death.

There is no more profound silence than that of the yew, a peculiar dense quality of quiet that seems to emanate from the tree itself, muffling and buffering the surrounding sound and separating those who stand beneath it from the rest of an overly noisy world. While other trees make a variety of susurrations in the wind – whisperings, rustlings, soughings and sibilant sighs – the yew seems not so much to produce sound as absorb it. I had experienced this distinctive quality of the tree when recording with Brother Sam under the canopy of the Tandridge yew, for though we were within the sound shadow of the nearby A22 and the not-too-distant M25, they somehow became an irrelevance. But we were struck, too, by another deep quality; that of its shade. The 'sable shade' Wordsworth had called it, beneath which he had imagined the gathering of noontide spirits:

> . . . *Fear and trembling Hope,*
> *Silence and Foresight; Death the skeleton*
> *And Time the shadow . . .*

The yew is the most shade tolerant tree in Europe, able to germinate where almost nothing else can grow, and it casts a deep, dense shade of its own. 'When men look upon the yew in wonder', wrote John Lowe in the introduction to his seminal book, 'there is in their minds none of the cheery, reverential feeling inspired by the "brave old oak" but rather a sense of awe produced by its sombre and gloomy foliage and the deathly character it bears in all its relationships.' Pliny's verdict was even more severe. 'Neither verdant nor graceful', he called it, 'but gloomy, terrible and sapless . . . unpleasant and fearful to look upon'.

The 'gloom' is not just in the intensity of its shade, it is in the very colour of its foliage, which is the darkest of greens; 'darker than night', as Geoffrey Grigson described it. Approaching a wooded hillside, it is instantly possible to pick out the yews, even from a distance, by their darker colour against the variable but lighter greens of the oak, ash or maple. Allen Meredith argues that such a dolorous view of the yew is a subjective product of conditioning and perhaps a symptom of the 'dislocation of the psyche'. 'This [funereal] impression changes', suggest Chetan and Brueton, in their book on his work, 'with a visit to one of the few yew groves in open country, such as Kingley Vale or Druids Grove, where the trees exist within their own silence and power.' I have visited both those sites and I cannot agree.

Kingley Vale, on the Chichester Downs, is indeed a beautiful place and I would gladly return there on a regular basis. The older yews grow as part of the mixed woodland that fills the hollow within the horseshoe shape of the downland, but higher on the flanks they form a continuous stand of young trees that extends for several miles, alleviated only by the bright, incongruous presence of a whitebeam. The darkness beneath these dense young yews is emphasised by the scattered white nodules of chalk that litter the otherwise bare ground. No plants grow and no birds sing. The words that came to mind were those from the Roman poet Lucan, quoted in rather contradictory fashion by Meredith himself: '. . . not a bird could be heard singing in this dark grove, nor animals would stay too long, neither Silvan priest would enter such a wood when sun had gone'.

No less shaded, 'Druids Grove' has a different atmosphere again. This is the name the Victorians gave to a concentration of yews within mixed woodland at Norbury Park in Surrey, on steep slopes above the valley of the River Mole. The yews here were older, straggling and spidery, and growing shoulder to shoulder with box, as they sometimes do in churchyards, the tall, dense bushes with their small, dull-green leaves only redoubling the tangled darkness. It was a stark and rather gloomy contrast with the sunny valley I had climbed from below, where colonies of large, edible apple snails lived on the woodland edge and ringlets, marbled whites and fritillary butterflies flew on the rides and fed on the white flowers of the brambles.

Plath's lines, I am sure, were about an emotional impact as much as a physical one and I should confess that, at times, I had become so immersed in the yew, as to feel I was being drawn back into darkness, the same darkness perhaps that, as a child, had drawn me to old churches and their symbols of mortality, and had afflicted me, at times in subsequent years, with bouts of depression. I had, I felt, wandered for too long in a place that was insubstantial, a place of shadow where fact and fantasy had become indissolubly merged.

But this is only one side of the story. There is another association of the yew that arises just as much from its physical characteristics and which seems to espouse the opposite. It is an association with immortality; not the immortality of eternal youth but that of eternal renewal, of continual dying to continually give rise to new life. This is the tree which Henry Vaughan, another metaphysical poet, described as,

A tree ne'er to be priced
A tree whose fruit is immortality.

In the old funerary practices the yew was not just marking a farewell to this earth but the opening of the passage to a new one. It was, of course, an aspect that Christianity was happy to incorporate. The Reverend C.A. Johns, whose *Forest Trees of Great Britain* was published by the Society for Promoting Christian Knowledge in 1849, developed the theme with vicarish zeal:

> *Heathens might indeed with propriety have selected the most deadly of trees to represent a merciless destroyer but such a feeling could have had no place with sober Christians. They, on the other hand, would regard the perpetual verdure which overshadowed the remains of their forefathers and was shortly destined to canopy their own as the most fitting expression of their faith in the immortality of the soul ... The yew, then, is ... an appropriate religious symbol.*

It fitted, too, with the theology of the Resurrection and in many places, especially Herefordshire, Worcestershire and Somerset, in addition to its processional use, the whole church would be decked with boughs of yew at Easter.

The idea that the yew can simultaneously stand for both death and immortality seems problematic to some. Thomas Browne wrote in *Hydriotaphia* in 1658: 'Whether or not the planting of yews in churchyards hold not its original from ancient Funerals or as an Emblem of Resurrection from its perpetual verdure, may also admit conjecture.' Some 350 years later, Andrew Morton expressed the same dilemma in his book *Trees of the Celtic Saints*: 'A question to be asked here', he said, 'is whether the foliage was a symbol of death or triumph over death'. The answer is both. Without death, immortality becomes meaningless and to see them as opposites rather than complementary aspects of a whole is, perhaps, to see the world too much through the lens of our monotheistic faiths. In folklore, especially the folklore of plants, contrary attributes regularly appear together; the hawthorn representing both love and death, the holly that can either protect or threaten, and the yew itself which as well as being poisonous is attributed with protective powers, and which as well as causing sickness provides a cure for cancer.

There is a recent award-winning children's book that has no problem with resolving the two, and I promised my niece Millie I would thank her in print for bringing it to my attention. In *A Monster Calls* by Patrick Ness, Conor's mother is dying of cancer and Conor, understandably, is unable to accept or come to terms with it. His mother often stands at their kitchen window gazing out at a yew tree that stands beyond the railway line across the valley. This yew begins to appear to Conor in monstrous form in his dreams, and relates a sequence of tales. Though terrifying at first they gradually help him to cope with his anger, to be alongside his dying mother and to accept that life will go on beyond her death. It is a moving tale, developed by Ness from an idea by Siobhan Dowd, who died of cancer herself before she had time to write it.

The yew as an attribute of immortality is, again, rooted in the characteristics of the tree, and the deep understanding that people would once have had of these. To begin with there is that 'perpetual verdure'. Since the globalisation of gardening and the importation of numerous new species, it is hard to realise the significance that evergreens once held in our landscape. The yew is one of only three native conifers, with the other two, the shrubby juniper and the tall Scots pine, being much more restricted in their distribution. There were hollies, of course, and a few shrubs like box and broom, but the yew would have been, then, one of the few countryside evergreens, and certainly the largest and most noticeable. Its presence, therefore, was all the more powerful, and its symbolism the more significant. Here was one of few visible sources of life-affirming greenery at this early time of year, a greenery which stood out against the general leafless greyness, and held out a promise, perhaps almost a prophecy, of the more general return of life to come.

In continental Europe, and in some Muslim countries, other evergreens have been linked to immortality, the cypress prominent among them. This, now, can be commonly found in British churchyards, along with other evergreens such as the box, the cedar and the holly. But beyond this shared display of continuous greenery, the yew has an aspect that is unique; it is the only tree that can continually renew itself. The yew has an amazing range of survival mechanisms, of which the ability to grow very slowly, or to cease growing at all, is one. It can produce adventitious growth in the form of buds, appearing from anywhere at all on its trunk. Its roots can sucker and produce new trees. It can grow vertical branches out of horizontal ones or from fallen lengths of trunk. As Brother Sam and I had seen at Tandridge, it can 'layer' itself, its spreading boughs arcing outwards and bowing to the ground and rooting where they touch the soil. I have seen these 'new', young trees clustering round a parent like a close-knit family circle, or persisting as a copse after the parent tree had gone. Or the arcing tree forms a huge vault, like a dark and empty cathedral; or takes on the more sinister form of a terrible terrestrial octopus. Even more remarkably, the decay of the heartwood and the hollowing out of the tree become, for the yew, not a mark of decline but just one of its stages of growth and it can encase and strengthen its thinning trunk with new and secondary growth. Inside that hollow, a root can grow down from the upper tree to reach the ground, a root that will swell and may eventually come to replace the old trunk and the 'old' tree.

This, then, is the death and life of the yew. And it was, I had become convinced, the real story behind its churchyard connection. In its darkness and silence, in its perpetual greenery, in its strategies for survival, in its great age and, above all, in its constant renewal, it became, for different cultures at different times, the symbol both of death, of life beyond it, and of the liminal space between. In Britain its association was with Neolithic burials and their barrows, and with early funerary customs that were common to much of the country. The arrival of the Romans, with similar practices of their own, would only have reinforced the connection as would that of the early Celtic saints, planting yews here and there outside their cells. In other places, the early Christians might have adopted for their church building a site, such as Fortingall, already marked by a tree of particular social or geographical significance. And so, gradually, the yew became the tree of the churchyard. Then the Normans arrived, and with them a new era of church building into which the yew is incorporated, seen as an equivalent, perhaps, of their own funerary cypress. Over time Christianity adds its own theology, and finds its own liturgical uses for the tree, and the connection is further strengthened. With its use for the national weapon, the yew becomes a national symbol, and an apt one for what becomes the national church. The chance discovery of a variant form on a Fermanagh hillside, gives us the Irish yew and adds to the story, for here now is a yew, abundantly planted, which is tidy and manageable and whose spires closely mimic the shape of the continental cypress. And so the tradition continues and is renewed today. The tree appears on a postage stamp and yew seedlings are distributed by the Conservation Foundation. Contemporary interest is further strengthened by our own vague yearning for some sort of 'reconnection' with nature. But it is, in essence, the yew itself that is the explanation for the churchyard yew.

And it still manages to manufacture mystery. Here is a tree that seems able to renew itself from inside out, meaning that none of its wood will be as old as the tree itself. It raises what seems an existential issue, one expressed brilliantly by Fred Hageneder of the Ancient Yew Group. The discovery of this capacity, he wrote, 'has forced science – or at least botany – to rethink its definition of "life" for plants. Is a tree that has completely renewed itself still the "same" tree? Is a living organism truly represented by its body or its genetic code or by something else? Is life more than what we can see and touch and measure?'

3

Star Fall

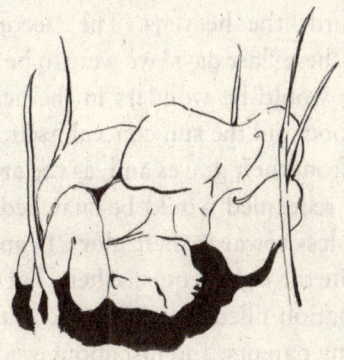

Things fall from the sky. From choirs of angels to frozen faeces from a passing plane, the air opens and they fly or drift or hurtle earthwards. They are the stuff of story and of science: meteorites and manna, red or black rain, fish and frogs in showers, thunderstones and gossamer, even the tortoise that killed the unfortunate Aeschylus. Their aerial provenance bestows an extra mystique, an additional layer of meaning. 'Up there' is the abode of angels, the hopeful location of heaven, the place where, in many cultures, the souls of loved ones have added their adornment to the bright multitude of stars. It is no wonder that in many languages the word for 'heaven' and the 'sky' are the same. But up there, too, within and beyond, is the eternal and ever-reaching darkness, the great cold; a fearful unknown emptiness that forms a final frontier. It is a place of threat, it is where the 'aliens' come from. Whatever falls from or through it, apparent or real, comes freighted with association; awesome, wondrous or fearsome, and sometimes all three.

In my fundamentalist upbringing, in a little Mission Hall between the bomb sites of Bermondsey, the eyes of the elders seemed perpetually to be trained skyward, like those of the contorted saints in Rococo paintings whose limbs twist into impossible positions and whose eyes strain perennially and agonisingly towards the heavens. The 'second coming' was imminent and in these 'last days' we were to be alert for signs of its arrival. There would be wonders in the heavens, the moon would turn to blood and the sun conceal itself in darkness. The dead would rise from their graves and, as the archangel sounded the trumpet, the redeemed would be snatched from the earth, and drift gravity-less towards their glory. I cannot vouch that I have events in quite the right sequence here, but there is no doubt that their anticipation filled my childhood with terror. I could clearly envisage my parents, and just about everyone else I knew, ascending to the rapture while I was abandoned below, the guilty secret of my unbelief revealed. Far later into life than I would care to admit, I could still be shaken from sleep by the tremor of a passing lorry, or by an ambulance blaring its siren in the street below, suddenly certain that, despite my doubts, the final trump was sounding.

The terror might have abated somewhat but the interest in things celestial was to be lifelong. In my teenage years it took the form of an enthusiasm for UFOs, or 'flying saucers' as we knew them, one that, to my relief, I shared with a number of my school friends. We attempted their study with what we considered a scientific detachment, though one in which the fascination was mingled with just a frisson of fear. We did, after all, know far too many stories of alien abduction. All of this was to come back to me when, many years later, poring over an eighteenth-century almanac, I first read accounts of another substance thought to have mysteriously materialised from space. Known as 'star jelly', accounts of it go back over a thousand years, and continue to appear, irregularly though repeatedly, today; but if its existence is well attested, its origin, and even its exact nature, is not. Star jelly remains shrouded in mystery, the subject of continuing and contested debate. Here was something else unsettling and said to have appeared from the skies. How, then, could I not be enthralled? The investigation into which it led me was to range from the most minute of organisms to events on a truly galactic scale; to questions about the origins of life on earth and the meaning of human intelligence; to cult horror movies and high Romantic fiction; to characters as diverse as Gulielma Lister and Paracelsus; and to new concepts which challenged conventional science. And then, just once, and when I was least expecting it, I was to encounter the substance myself.

* * *

Whatever the journey of discovery it had led me on, the physical appearance of star jelly is not one of immediate appeal. It takes the form of gelatinous or slimy lumps, sometimes white and opaque, sometimes green, brown or purple, and sometimes clear and transparent. It may appear as a single quivering globule, or sometimes as many. It can be odourless or emit a foul smell. Usually just a matter of centimetres across it has also been described in masses of a metre or more. Our historical ambivalence towards it is revealed in the variety of names we have given it. As well as 'star jelly', it is 'star shot', 'star fall', 'sky', or the pseudo-scientific 'astromyxin'. It is also 'star-rot', 'star-slime', 'star-slough' and even 'star shit'.

Star shot or star shit, what all these names reflect is the earliest theory of all about the origin of star jelly, and the one that has been most persistent; that it is the product of shooting stars. It is an idea that dates back to at least the seventh century and the writing of the Venerable Bede. Setting out a theory for the formation of what he called 'falling stars', he suggested that they were sparks, caused by the collision of rushing winds. Extinguished by damp air as they fell to earth, they appeared in the fields as masses of 'poisonous phlegm'. The idea of a stellar origin for the jelly-like substance was to stick, but not everyone had such a disparaging view of its value. Writing to Prince Henry in the early seventeenth century, Francis Bacon, now widely considered to be the inventor of the scientific method, described it as a 'magical ingredient' and, over the centuries, the alchemists came to value it highly. What better substance, after all, than a seemingly celestial jelly, to assist in the search for the elixir of immortality, for the *alkahest* or universal solvent, for the *menstruum*, the wonderful substance that would transmute base metals into gold?

The physicians took a similarly positive view. One of the earliest written references after Bede comes from the work of John of Gaddesden, a contemporary of Chaucer who drew his name from the village of Little Gaddesden in Hertfordshire. *The Rosa Medicinae*, his medical compilation of 1305, recommends it for the treatment of abscesses, describing it as *'stella terrae'*, 'a certain mucilaginous substance found lying upon the earth'. Other texts, from both England and France, suggest its use in the treatment of cancers, fistulas and hair loss and, on a more veterinary note, for the soothing of sore heels in horses. In 1661, Robert Boyle, a founder member of the Royal Society best remembered by students for his law on the volume of gases, described another medical use:

I have seen a good quantity of that jelly that is sometimes found upon the ground, and by the vulgar called a star-shoot, as if it remained upon extinction of a falling star, which being brought to an eminent physician of my acquaintance, he lightly digested it in a well-stopt glass for a long time, and by that

means alone resolved it into a permanent liquor, which he
extols as a specific to be outwardly applied against wens.

There is a note of disdain in Boyle's dismissal of the 'vulgar', a
note that recurs in scientific writing on the subject and reveals,
perhaps, a too ready willingness to reject the accounts of ordinary people. Nonetheless the 'vulgar' remained sure of it; the
substance had fallen from the sky.

Sometimes it appeared in daytime, seeming to have come
from the clouds. There is a Welsh account of a labourer who felt
it fall on his head as he worked in the fields, and in the
Philosophical Transactions of 1756, W. Watson relates how 'our
country people call it [s]tar-slough and some of them, as it is
principally seen after rain suppose ... that it drops from the
clouds'. More commonly it continued to be associated with the
night sky and with shooting stars. Names suggesting a stellar
origin had, in fact, been around since at least 1440 when the
word 'sterre slyme' appeared in the *Sinonima Bartolomei*, an
English–Latin dictionary. Such names were by no means
restricted to English. The Welsh had *pwdre ser*, which, noble
though it sounds, means 'rot of the stars', while other equiva-
lents appeared in Dutch, Danish, Swedish and Frisian. The
Germans had several words for it, including *meteorgallerte* for
meteor jelly and, less politely, *sternenrotz* or star snot. The
French were alone in following a different scatological route. To
them it was *le caca de lune*, or 'moon poo'.

* * *

They are unflattering names for the possible product of that
magnificent brief brightness seen streaking across an open night
sky. Today we would know them as meteorites, and far more
solid in their substance, but their lexicography remains almost
as diverse as that of the jelly. They are 'meteoroids' while still in
deep space, 'meteors' when they reach our atmosphere, and

'meteorites', or even 'bolides', when they hit the ground. They may not now have alchemical associations but they are awe-inspiring in their own right, and increasingly so the more we learn about them. They are also with us in far greater quantity than we might realise. According to one recent estimate, a hundred metric tons of cosmic material enters our atmosphere every day, enough to provide a particle for every square metre on earth. The vast bulk of this arrives from the asteroid belt but it can also consist of matter from the tails and trails of comets, or of fragments thrown up by collision impacts on the surface of the moon or Mars. Entering the atmosphere at speeds measured in tens of thousands of miles an hour, and now subject to friction, pressure and new chemical interactions, they heat up and begin to radiate energy. I have seen them only as sudden short streaks, the 'sparks' of the Venerable Bede, but sometimes their fireballs can rival the sun in intensity, burning yellow, red or green with flashes or bursts of light as the matter within them explodes or disintegrates. They can give rise to detonations, rumblings, strange whistling or hissing sounds, or sonic booms heard for a radius of sixty miles or more. It is fortunate for us that only 5% of such matter reaches the earth's surface, much of it falling as dust. Those fragments of any size that do land among us have an impact that is both physical and cultural.

Every year in the season of Hajj, more than two and a half million pilgrims circle the Kaaba in the holy city of Mecca. It is a cube-shaped building, draped in black cloth with a frieze of gold embroidery. Set in its eastern corner are the several fragments of the 'Black Stone', cemented together within a silver frame. It is the 'al-Hajaru al-Aswad' which marks the point where the seven required circuits of the Kaaba begin. Its significance as a sacred object actually predates Islam and the question of its origin has been the subject of much debate. It is, by popular belief, the remains of a meteorite, though one recent theory amends this by suggesting it is an impactite, the stone from around a meteorite, turned vitreous by the heat and pressure of its arrival. In Islamic tradition the Black Stone fell from heaven in the very earliest times and indicated to Adam and Eve where they should build the first altar. As the pilgrims pass by on each circuit, they press forward to kiss or touch it or simply to point towards it, for it is the prophet Muhammad himself who is believed to have set it there.

Among other sacred references to meteorites is the Temple of Artemis, one of the 'seven wonders of the ancient world', said to have been built on the site where a meteor landed, and believed to be a direct message from Jupiter. It was an idea echoed by the film director Werner Herzog when he described meteors as 'e-mails from God', though I would personally prefer an image more powerful than that of a deity on a laptop. Reverential attitudes towards the meteorite are entirely appropriate for such a significant and otherworldly object. They echo the alchemical treatment of their supposed product as a 'magical substance'. They may not have bestowed immortality but they are, we now know, the bringers of both death and life. They predate the formation of the planet itself and their impacts have changed evolutionary history. And found within them, if not the *alkahest* and the *menstruum*, is a structure found nowhere else on earth, and until recently, believed to be impossible.

* * *

Until the 1980s the existence of a five-fold symmetry in crystals was held to be against the laws of science. Crystals could exist in cubes and hexagons and other repeated patterns, but a 'quasi-crystal', one with an irregular number of sides, was a logical impossibility. The rather prosaic analogy usually applied was that of laying bathroom tiles; for if such tiles were five or ten sided there would always be gaps appearing between them. A quasi-crystalline structure, with regular but non-repeating patterns, represented a blending of periodicity and irrationality that was anathema to scientific thought. When, in 1982, the Israeli scientist Dan Shechtman demonstrated its theoretical possibility, he was met with hostility and ridicule. He was asked to leave his research team for bringing it into disgrace, and the double Nobel laureate, Linus Pauling, pronounced witheringly that, 'there are no such things as quasi-crystals, only quasi-scientists'. Though it took Shechtman another two years to publish his findings, they were to lead in 2011 to his Nobel Prize for Chemistry. His work, said the awarding committee, 'forced scientists to reconsider their conception of the very nature of matter'.

Once the possibility of quasi-crystals had been established, it did not take long for researchers to produce them in the laboratory, but it was a few more years before they were to be found in nature. Then, in 2009, working on samples taken from the Khatyrka meteorite in north-eastern Russia, a small team of researchers found three quasi-crystals. At 4.57 billion years old, the Khatyrka meteorite dates from the formation of the solar system itself; from a time when the cloud and dust of the solar nebula was beginning to aggregate into particles, and when the colossal forces of heat and energy then at work could combine to produce this 'forbidden symmetry'. The impossible crystals were shown to exist – the result of an astral alchemy.

As the oldest material to be found on earth, meteorites are helping us unravel the history of the universe, but they do not just interpret planetary history, they have also played a significant part in shaping it. Somewhere around 65 million years ago, a meteor more than six miles across and travelling at twelve miles a second, hit the earth with a force equal to that of 10 billion of the bombs that devastated Hiroshima. Incandescent material ignited huge wild fires as far as 900 miles away, while colossal shock waves triggered earthquakes and volcanoes on a global scale. Tsunamis, with waves nearly a mile high, reached every coastal region on the planet while from a crater nearly a hundred miles in diameter, rose great clouds of hot dust, ash and steam. The resulting mass death across a vast area is still visible in the fossil record, but it was only the beginning. The dust and particles rising from the impact site, on what is now the Yucatan peninsula of Mexico, formed a dense covering of clouds, cocooning the earth and blocking the sun for perhaps as long as a decade. With the amount of sunlight reduced by more than 50%, with an increase in acid rain, the acidification of oceans, a cooling of the earth's surface and a reduction of photosynthesis in plants and plankton, the scene was set for a global ecological disaster. The Chicxulub meteor is now generally accepted to have triggered one of the earth's six great extinction events, a view confirmed in 2020 by a panel of forty-one experts reviewing twenty years' worth of data. The K/T or Cretaceous–Tertiary extinction removed 75% of species on the planet, including the ammonites, many flowering plants and, most famously, all of the non-avian dinosaurs. In each of the mass extinctions it is the largest and dominant species that are most affected, and their removal provides new opportunities for evolution and for diversification. So it was with the K/T event. Up until then the mammals had been a relatively small and insignificant group. But they survived, and with their main competitors removed, rose to prominence. With them, for better or worse, came *Homo sapiens*. We are, it could be said, one of the impacts of a meteorite.

* * *

The possibility of another Chicxulub is still with us. According to the North American Space Agency we are surrounded by more than 19,000 'near earth objects', roughly half of them larger than 140 metres. Around thirty more are being discovered each week. It was to monitor the possibility of another catastrophic strike that NASA established its Near-Earth Object Observations Program and a Planetary Defense Coordination

Office. But if a meteor could end life on earth, there is also a possibility that it brought it. Around 5% of meteorites are of a type called carbonaceous chondrites and they contain the organic compounds that provide the building blocks of life: ribose and bioessential sugars, lipids and hydrocarbons, alcohols and complex amino acids. Overall eighty types of amino acid have been found in meteorites, with a single meteorite, the 1969 Murchison meteorite, containing over a dozen of them. Studies in 2018, applying new techniques to meteorites that had fallen some twenty years earlier, revealed that they contained not just amino acids but liquid water, one of the other prebiotic essentials. Meteorites might even have conveyed already living bacteria onto the planet. DNA is robust enough to survive in deep space, whilst many forms of bacteria are found in rock. In fact some 5–15% of all our biomass lives within the earth's crust, including 90% of all bacteria, some as much as a mile or more beneath the surface. Alternatively, meteorites may have conveyed to earth all the ingredients for life, fortuitously then finding a suitable environment for its emergence.

Writing in the journal *Astrobiology*, in September 2020, a group of Canadian scientists went even further. They argued that meteorites might not just have delivered the ingredients for the generation of life, but may also themselves have created the conditions under which it could flourish. 'Perhaps unsurprisingly,' they write, 'there has been a general tendency in past decades to think of impacts as primarily destructive events that would have endangered life on early Earth.' They call instead for 'a paradigm shift in our view of the biological consequences of meteor impact events . . . (which) are not just isolated catastrophic geological events but a fundamental process in planetary evolution that plays an important role in the origin of life and in controlling planetary habitability.'

Meteorite strikes might have contributed to this 'habitability' in a number of ways. The shock heating of impacts within a primitive atmosphere could itself produce additional materials such as hydrogen cyanide, important in the formation of prebiotic molecules. Impacts, such as that at Chicxulub, might also have played a role in creating habitats highly conducive to microbial colonisation, such as hydrothermal vents, hot springs, volcanic splash pools and crater lakes. 'Would it not be poetic,' the article concludes, 'that impacts long seen as harbingers of death, turn out to have in fact been the cradle of life?' Poetic indeed. The meteorite giveth and it taketh away. But the paradox of planetary events that are both cataclysmic and creative, should not surprise us. They occur throughout nature. Fire, flood, earthquake and volcano are all part of a wider pattern, wonderfully terrifying, but essential to life on earth, and without them the earth would be barren. It was volcanic eruptions as a source of carbon dioxide in the atmosphere, warming the planet and protecting it against UV rays that helped make life possible. It was the drifting of the cooling planetary crust and the crashing and grinding of plate tectonics as they threw up our mountain systems that determined and distributed our rainfall. It is the continuation of that building and erosion of mountains that maintains the supply of nutrient rich soils, and it is the continuing volcanoes that cycle to the surface huge quantities of the minerals essential for life, just as floods still redistribute them as fertile soils. It is the dreadful human dilemma; that the very forces that threaten our lives are those that also make it possible. Hinduism has recognised this paradox for thousands of years in the form of Shiva, the great and terrible god, creator and destroyer, full of contradictory elements, who sweeps away the illusions and imperfections of the world, paving the way for beneficial change. 'In my end is my beginning', as T.S. Eliot put it, and 'In my beginning is my end.' But the cycles at work here are not just bigger than human life; they are bigger than the life of the planet.

* * *

Though star jelly has never been found alongside any of the more solid matter associated with meteorites, references to a relationship between star jelly and falling stars continued throughout the seventeenth century. When Robert Fludd, a physician who also had an interest in the occult, saw a meteorite fall in 1619, he went to the site as early as he could the next morning and found what he took to be its remains; a mass of white slippery substance bearing small black spots. Other commentators, whilst maintaining a personal scepticism, describe how this celestial origin remained a common belief. 'Amongst ourselves,' wrote Thomas White in 1656, 'when any such matter is found in the fields, the very countrymen cry out it fell from heav'n and the starres, and as I remember, call it the spittle of the starres'. The same idea was still being expressed the following century by John Morton in his 1712 *Natural History of Northamptonshire*. 'I shall here set down', he writes, 'my Remarks upon the gelatinous Body call'd Star-gelly, Star-shot or Star-fall'n, so named because vulgarly believ'd to fall from a Star, or to be the Recrement of the Meteor which is called the Falling or Shooting Star.'

Perhaps there was another way in which those vulgar 'countrymen' were envisaging the stars as fallen to earth. If the popular names of our wild flowers are to be believed, they could be encountered almost everywhere. They were there in the star-flower, the starwort, the star grass and the star-thistle; in star-fruit, starlight, star-reed and star-of-the-wood, and even in the enigmatic star-naked boys. In all, the word 'star' appears at least thirty-nine times in our regional plant names. The more specific star-of-Bethlehem appears three times, including the beautiful garden plant and wild flower *Ornithogalum umbellatum*. I found it once, growing in thick grass in the ancient graveyard at Iona Abbey. When I took friends to show them the next day, a tidy gardener had strimmed it away. Stars appear again in the scientific names for our plants, from the two different Latin words for star. *Astra* gives us the asters, perhaps better known as the Michaelmas daisies. Originally garden plants, they are now often rampant in the wild. I have seen their purple flowers in magnificent autumnal masses over Walthamstow Marshes. I cannot say that I find them particularly star-like but they do have something of the darkness of the firmament about them. From *stella* we get another genus, the *Stellaria*. It includes both the sprawling, pavement chickweeds and the delicate woodland stitchworts, whose petals are so deeply notched they look like two separate rays. In April their carpets of white flowers light up the early grey and fully earn them their stellar designation. Perhaps the most starry plant of all, however, is the common daisy. Though its familiar name derives from the 'day's eye' of the sun, it counts both 'star' and 'little star' among its local variants. I have seen it flowering so thickly in a Hebridean field that it seemed an earthly reflection of a densely starry night sky. It must have been a sight like this that led Thomas Hardy to talk of the 'constellated daisies', or the Scottish poet and clergyman, Andrew Young, to liken them to the 'powdery light' of a starry sky glimmering 'from the night-strewn ground'.

Even the fungi have their own stars; a whole family of them, in fact. Emerging in a ball shape, the Geastrales, or earth stars, have thick cases that split open and peel back to form anything from four to twelve rather stubby rays. Cupped within them is a softer pore sac and often the tips of the open rays will curl backwards and beneath themselves, arching the sac upwards and off the ground, as if in some awkward yogic position. When I last found a troupe of the collared earthstar, *Geastrum triplex*, growing beneath a privet hedge on the outskirts of Brighton, they resembled not so much a constellation of stars as a troupe of multi-legged, crab-like creatures scuttling for cover beneath the foliage; or perhaps the landed pods of an alien spacecraft, as in the film *E.T.* It might not be too fantastical a thought. In 1847, the mycologist C.D. Badham suggested that *Geastrum*, 'aspiring occasionally to leave this earth had been found suspended . . . between it and the stars, on the very highest pinnacle of St. Paul's'. Someone called Withering, he explained in a footnote, 'found one of these on the top of St. Paul's Cathedral; the first he had seen!' Like Withering, I have been to the top of St Paul's, and from it I have seen peregrine falcons wheeling above the City and coming to rest on the tower blocks; but I have not seen anywhere up there that an earth-star might find purchase.

The fungi are the only group, moreover, to include not just stars but a shooting star. *Sphaerobolus stellatus*, the shooting star fungus, is not uncommon and is found in the unglamorous setting of rotting wood, sawdust heaps, old sacking, dead leaves, damp sticks and dung. Nor is it prepossessing in its appearance. It grows in masses of globose, whitish spheres, only a couple of millimetres across, which split open into five to nine tiny orange rays; not so much like stars as party hats from a miniscule Christmas cracker. Contained within each is a small, brown ball-shaped spore sac. When conditions are right, a lining of the rayed case suddenly everts itself, pushing upwards and turning itself inside out in something like one thousandth of a second, and with such force that it ejects the spore sac into the air. The distance covered by these 'shooting stars' is remarkable. Writing in 1909, the mycologist Arthur Buller described the process as something of a field sport, like throwing the discus or putting the shot, for which he recorded 'Olympic records'. The spore sacs, he said, could be cast as high as 4.3 metres, and land at a distance of more than 5.5 metres. Given the tiny size of the fungus, this was, he suggested, equivalent to a six-foot man throwing a baseball one and a half miles into the sky and landing as much as two miles away.

* * *

Implicit in all these starry names for flowers and fungi is an easy familiarity with the night sky and its components. They suggest a very different experience from that available to most of us today. Dusk has fallen as I write this and as soon as night gathers itself sufficiently, I step out into my back garden. Here, in the East End of London, the plane trees form stark winter silhouettes and above them the sky has a patchy lining of thin grey cirrus clouds which seem to be illumined with reflected light from beneath. The moon is as light as a balloon and icebright. It is three quarters full and waxing, in the phase that goes by the strange name of gibbous. Close to it is a single steady point of light which, from its red tinge, I know to be Mars. Further away, and threatening to disappear behind the church tower, is another point of light, paler and more yellow. I am guessing it is Venus, the beautiful planet that can be either the evening or the morning star. Then I count the true stars. It doesn't take long. There is only one of them visible. It must be the Dog Star or Sirius, named from the Greek word for 'glowing', which it must be to survive the onslaught of ambient light that has driven every other star into hiding. It is if I am viewing the sky through a cataract, a cloudy veil that dims its detail. Dazzling security lights burn night-long outside our parade of small shops, white lamp-shine suffuses the surrounding streets and the estates are a patchwork of brightly lit windows. Plagued by light, the city sleeps but fitfully. Robins sing by midnight lamplight and foxes haunt the sleep of bins. Our especial curse are the high glass-and-steel towers of Canary Wharf, the financial district that overlooks us. They act as an exact converse to the natural order of things; darkening us with shadow by day and shining out on us by night from forty floors of lit-up office windows. As new blocks go up on our side of the tracks they have been accompanied by a marketing billboard: 'Why move to Canary Wharf when you can watch it light up from here?' Who needs, it seems to suggest, the old and tired panoply of night and day when we have constructed our own bright, new alternative? With over 84% of the United Kingdom population now living in cities, how many of us have actually seen a shooting or falling star, or experienced the thrill of its brief, bright passing?

* * *

Accounts connecting star jelly to meteorites were still common in the nineteenth century. The 30th annual meeting of the British Association for the Advancement of Science, for example, reported on a 'shooting star' falling to earth in Silesia in January 1803. Its trajectory was low, with witnesses hearing a 'whizzing sound' as it passed overhead. It was still burning when it hit the ground, allowing its point of impact to be easily observed and investigators in the morning found there, 'a mass of jelly like substance'. Similarly the *Edinburgh Philosophical Journal* of October 1819 reports on a 'fiery globe' that had fallen on the island of Lethy in India. Searchers who went to the site found a large quantity of gelatinous material, which quickly disappeared. Three years later, according to the *Annual Register* of 1821, an object was seen to fall with a bright light in Amhurst, Massachusetts. Foul smelling and covered with a cloth-like nap, it was examined by Professor Russell Grove who removed the surface to find a buff-coloured pulpy substance beneath. In January 1834, *The American Journal of Science and Arts* included a 'Report in Connection with a Meteor Shower over Eastern USA'. The people of Rahway, New Jersey, had seen a 'fiery rain' fall to the ground, depositing lumps of a dark, yellow, foul-smelling jelly. On 11 November 1846, according to the *Scientific American*, an object that fell in Louisville, New York State, 'appeared larger than the sun, illumined by the hemisphere nearly as light as day . . . a large number of the citizens immediately repaired to the spot and found a body of fetid jelly, four foot in diameter'. Such reports continue throughout the century, and from places as far apart as Lithuania in 1846, Genoa in 1870 and Turkey in 1891.

Not everyone, however, was accepting of a meteoric origin for the jelly. Among them was the novelist Walter Scott who, writing in 1808, rejected the idea as something of a childish delusion. Editing a volume of the works of John Dryden, he uses one of his footnotes to comment that:

> *It is a common idea that falling stars, as they are called, are converted into a sort of jelly. Among the rest, I had often the opportunity to see the shooting of the stars from place to place, and sometimes they appeared as if falling to the ground, where I once or twice found a white jelly-like matter among the grass, which I imagined to be distilled from them; and thence foolishly conjectured that the stars themselves must certainly consist of a like substance.*

His youthful encounter did, however, make some sort of impression for he refers to star jelly in his 1825 novel *The Talisman*.

It is one of his Waverley novels, set at the time of the Crusades, with Richard I joining other European rulers in an attempt to retake Jerusalem. Scott clearly had a sceptical view of the enterprise, depicting the sordid scheming, the struggling for precedence and the narcissism that divided the participating forces. As its antithesis he presents an idealised view of a chivalric code, transcending political and religious divides. Richard, despite his irascibility, is clearly an exemplar, but it reaches its epitome in the refined and civilised Saladin, the leader of the Muslim forces. He is, perhaps, the least unlikeable character in a book whose wooden theatricality makes it hard work for a modern reader. It abounds with long, florid and circumlocutory sentences, and there is not a line of dialogue that is not prolonged with elaborate formalities, formulaic courtesies, adornments, elaborations and a dash of moral exposition. Read one sentence in three, I began to think, and I would know just as well where the story was going.

It is not until Chapter 28 that the reference to star jelly appears. Theodoric of Engaddi, an enigmatic central character, is relating his back-story to King Richard. It is a sort of moral warning; he was once of high position but amorous misadventure has led him to his present humble state as a half-mad hermit subsisting in a desert cave. 'Seek a fallen star', he says, 'and thou shalt only light upon some foul jelly, which, in shooting through the horizon, has assumed for a moment an appearance of splendour'. Scott's use of the image as a metaphor for disillusion is instructive. Probably he had come across it first when he was editing Dryden, for it appears twice in this way in the work of the Restoration poet and dramatist. 'The shooting stars end all in purple jelly', he had written in his 1678 play *Oedipus*, and returned to the idea the following year in *The Spanish Friar*: 'When I had taken up what I supposed to be a fallen star, I found I had been cozened with a jelly'.

Since Perry Como sang 'Catch a Falling Star', and advised putting it in your pocket to save for a rainy day, we have had a more romantic and Disneyesque association with the phenomenon, but in earlier literature, the falling star, and the jelly associated with it, were emblematic of failed promise and of disappointed hope. It was a particularly popular device among the metaphysical poets, with their love of an extended conceit, and appears in the work of Richard Lovelace and Abraham Cowley as well as of John Donne. His 1613 'Epithalamion', celebrating the scandalous marriage of Frances Howard to the Earl of Somerset, describes the bridegroom's arrival:

> As he that sees a star fall, runs apace,
> And finds a gellie in its place.

The most bitter reference, however, is in the work of John Suckling and his poem 'Farewell to Love'. Here, having been let down in love, he rather petulantly renounces the emotion completely:

> . . . *my dear nothings, take your leave;*
> *No longer must you me deceive,*
> *Since I perceive*
> *All the deceit, and know*
> *Whence the mistake did grow.*
> *As he whose quicker eye doth trace*
> *A false star shot to a marked place*
> *Does run apace,*
> *And thinking it to catch*
> *A jelly up does snatch*

Now, in typically misogynistic fashion, he sees not women, but corpses, and castigates his eyes as the source of his folly, concluding,

> *These of my sins the glasses be:*
> *And here I see*
> *How I have loved before.*
> *And so I love no more.*

* * *

John Dryden, the metaphysicals, and a few mid-eighteenth-century poets were not to be the final cultural appropriators of the jelly-filled meteor and it returned in the mid-twentieth century in very different form. In a widely reported case in September 1950, police officers John Collins and Joe Keenan were driving their patrol car through the streets of Philadelphia when they saw a strange, shimmering object coming to earth in a field nearby. Going to investigate, they found a domed disc of jelly emitting a purple glow. When they attempted to pick it up, it fell apart leaving fragments stuck to their hands, and within an hour the whole mass had evaporated. It was this incident that was said to have inspired the 1958 schlock horror movie *The Blob*, a film that was to become something of a cult.

Directed by Irvin S. Yeaworth Junior, it featured Steve McQueen in his first leading role and was distributed as a double feature along with *I Married a Monster from Outer Space*. It had originally been called *The Molten Meteor* until its producers overheard the screenwriters refer to the film's malignant entity as 'the blob' and adopted this as its title instead. It was, said Arthur Halliwell in his review, 'padded hokum for drive-ins' and, indeed, it can be seen playing in the background in the drive-in movie scene in *Grease*. It is set in an idealised small-town America, where elderly men are referred to as 'old-timers', the girls have narrow waists and spreading skirts, and the boys are respectful and cruise in open-top cars. A meteorite lands just outside the town and breaks open to reveal a congealed jelly-like substance within. It adheres to the hands of the man who first uncovers it, and who injudiciously pokes it with a stick, and then proceeds to consume him. Growing in size, it sets out in a determined but unspecific way around town, devouring whichever unfortunate comes into its path. The clean-cut courting couple, played by McQueen and Aneta Corsaut, were witness to its arrival and it becomes their job to convince a sceptical police force, and the rest of an almost terminally supine town, of the existential threat about to engulf them.

The music for the film was written by Ralph Carmichael but a title song was added by Burt Bacharach in one of his earliest ventures into screen music. Now part of the cult appeal, it is oddly jaunty for a horror film, sounding more like an advert for a stain remover than a curtain-raiser for an extra-terrestrial all-devouring jelly. In what is probably the film's most iconic scene, the Blob, having ingested the projectionist at the local cinema, squeezes its way into the packed auditorium, threatening the entire, and largely teenage, audience. An exterior shot shows hundreds of terrified young people spilling out, screaming, onto the street. Since the year 2000 this scene has been re-enacted with locals as part of an annual 'Blobfest', held in Phoenixville,

Pennsylvania, one of the towns in which the film was made. Easy to find on YouTube, it is hugely enjoyable, with the crowd, some in completely unconnected monster costumes, waving their arms as they emerge, and attempting either a look of terror or a tentative scream. They are almost equally divided between those appearing embarrassed and those embracing the experience with enthusiasm. But just in case there is too much enthusiasm, a supervising safety marshal corrals them all in the correct direction. It is horror, but horror with health and safety.

The Blob was to see a sequel in 1971 and a remake in 1988. Directed by Chuck Russell, perhaps best known for *The Mask*, the intention, no doubt, was to take advantage of more recent advances in special effects. The result, despite a much larger budget, does not discernibly increase the horror quotient. What does become interesting, however, is the sociological difference between the earlier and the later versions. The original *Blob* was made just a few years after the McCarthy era, at a time when the red scare and the fear of international communism was still at its height. Some commentators have interpreted the 1958 version as a parable on the inexorable advance of this un-American menace. The jelly in the film was, after all, coloured red, and its final defeat by 'freezing' could be seen as a reference to the Cold War. One might also cite as circumstantial evidence, that the makers of the film, including the director and the screenwriter, were conservative evangelical Christians who had previously only made religious films, including promotions for the evangelist Billy Graham. Come 1988, however, and there are important changes in the remake. The town is somewhat seedier – it has clearly seen the effects of recession – and the youth more dissolute and less respectful. More significantly, the central tenet of the plot has changed. The jelly still arrives in a meteorite, and is still intent on its devouring mission, but the source of the threat is different. In fact it is internal rather than external. The Blob has no longer arrived from a distant planet; it is instead, we learn, the rogue result of home-grown experiments in biological warfare. It is the scientists in the remake who become the villains, trying to conceal the origin of the threat and their responsibility for it, and showing themselves prepared to sacrifice an entire town in order to keep their secret. A respect for the authorities has been replaced by suspicion; and a concern for the creeping power of communism, by a fear of the destructive possibilities of science.

Though already defeated three times by 1988, the Blob survived. And not only survived but went into politics. 'The Blob' was the name used by William Bennett, US Education Secretary in the 1980s, to deride bureaucrats, unions and educational researchers sceptical of his attempted reforms. From here the term was to enter the lexicon of the American libertarian right, promulgated through their network of think tanks and applied to the defence establishment as well as to any other grouping that seemed oppositional to their ideals of individual autonomy, market supremacy, low regulation and a small state. One of the British adherents to this political philosophy was Michael Gove, founder of his own right-wing think tank, the Policy Exchange. It was not just the ideas that he imported but also the language. During his time as Secretary of State for Education, assisted by his political adviser Dominic Cummings, he railed against what he called the 'Educational Blob' of teachers, officials and academics who resisted his ideas. As the Rasputin of British politics, Cummings went on to become chief adviser to the Prime Minister, extending his analysis of the Blob to the entire Civil Service, which he famously advertised to replace with 'weirdos and misfits'. With his sudden and ignominious departure from Downing Street in November 2020, a supporter wrote disconsolately to the *Sunday Times*. 'The Blob,' he said, 'has won'.

* * *

Throughout all the reports of meteorites and star jelly, continuing into the twentieth century, there had always been a contrary strand of scepticism in which Boyle and Scott were not alone. As new theories began to emerge, or were resurrected, *The Gentleman's Magazine* was once again on the front line of the debate. A correspondence in its pages was launched by 'MS' of Shotton in Flintshire, in a letter dated November 1774. The microfiche copy I consulted was damaged, its print overlain by that from another page that, at some stage, had been pressed against it. It took me some time to decipher, and even then various words and phrases remained illegible:

> *Mr. Urban,*
>
> *In a late(?) tour I happened to meet with what I thought was a curious phenomenon, tho' indeed I have often heard of such like matter and, if I mistake not, have seen it before . . . The matter resembles jelly or glue when wet (?) . . . it is translucid and the least touch puts it into a tremble . . . The field where I found it was pasture, had then cattle in and is pretty dry being a . . . [illegible] . . . The owner of the field being . . . [illegible] . . . old gentleman was very careful in preserving it and in remarking the place where it lay, in order to find out whether any alteration is made in the future produce.*
>
> *And now, Sir, having said all that I intend on the product of my evening walk, and not finding anything of the kind mentioned in any book that has come into my hands wherein it might be expected; I hereby request the favour of some kind reader, that he will inform me either what he has seen and where, or of his opinion concerning such matter and he will greatly oblige your humble servant.*

Several kind readers did indeed oblige. The first response came from 'RP' in March 1776 and was headed, 'Curious account of the Jelly-like Substance found in Fields':

> *I take the liberty of giving your correspondent some informa-*
> *tion relating to the jelly-like substance he enquires after . . .*
> *The substance is not unfrequent in England, nor in all the other*
> *parts of Europe, after rains, both in spring and autumn. Very*
> *large spots of it are seen in gravelly soils, and particularly on*
> *the tops of hills, and on open downs, and often it is found on*
> *gravel walks. It is met with in some of the old authors under*
> *the name of Nostoc and in Paracelsus . . .*

The word 'Nostoc', or sometimes 'Nostoch', appears several times in the correspondence, usually coupled with the name of Paracelsus. Perhaps less known now than he should be, he was an important, contradictory and controversial figure in the history of science; and significant, too, in the story of star jelly. Born in 1493, near Zurich, his given name was Philippus Aureolus Theophrastus Bombastus von Hohenheim, which might explain why he adopted a pseudonym. His choice for one was significant. Philosopher, physician, astrologer and alchemist, he was also a life-long iconoclast, detesting the rigid authority of the universities of his day and railing against their dependence on texts by authors who had been dead for a thousand years or more. The 'Paracelsus' pseudonym, first used in 1529, literally means equal to, or surpassing, Celsus, one of the revered authorities that he despised. His belief was in the primacy of observation over book learning. 'The patients are your textbook,' he taught, 'the sickbed is your study'. In a notorious incident he symbolically burnt volumes by Galen and Avicenna at one of the students' traditional midsummer fires. Wanting to be accessible to all, always an unpopular approach with the professions, he often lectured or wrote in his native German rather than the usual Latin. Travelling extensively in Europe, and beyond, he collected folklore and folk medicine and sought out the knowledge of peasants, healers and craftspeople. His preparedness to learn from the experience of ordinary men and women, marks him out from the scholars of both his day and ours.

His life was as outrageous as his theories. He was wild, brilliant and mercurial, prone to abusive outbursts and despising the titled people of his time. He was renowned for drinking with his students and for his numerous relationships with women. His enemies were many and he was pursued both inside the courts and out, including surviving several death threats. Opposed by conventional physicians, he was barred from practice wherever he attempted to settle and several of his publications were banned. Dismissed by the establishment, he was, nonetheless, revered by many and at one time likened to a medical equivalent of his contemporary, Martin Luther, a claim that he disowned. He died in 1541, at the age of 48, whilst staying at the White Horse Inn in Salzburg. In a strange parallel with David Douglas, the circumstances surrounding his death remain mysterious. It followed a very brief illness which some believed to be the results of a poison attempt, others arguing that it was a delayed result of his own experimentation with mercury. Either way it occurred shortly after a scuffle, said to have been with 'assassins', possibly in the pay of the orthodox medical faculty, during the course of which he was thrown down a hill. It almost certainly hastened his demise.

Controversy continued to pursue Paracelsus after his death and for several centuries the historians of science treated him dismissively. His last book, *Astronomia Magna*, published posthumously, contained his ideas on astrology, divination, theology and demonology, and these, as well as his prophecies, were taken up by secret societies like the Rosicrucians. Neither will his reputation have been helped by the contemporary disdain for the alchemists, as having something of the occult about them. In his seminal TV series, and later book, *The Ascent of Man*, Jacob Bronowski took a different view. 'Alchemy', he argues, 'is much more than a set of mechanical tricks or a vague belief in sympathetic magic. It is from the outset a theory of how the world is related to human life.' If it seems childish to us now, he continues, how much more 'our chemistry will seem childish 500 years from now . . . A theory in its day helps sort out the problems of its day.'

Paracelsus, in Bronowski's account, becomes not a quack but a 'profound and divided' genius. It was part of a much wider rehabilitation of a man whose achievements were enormous. Introducing minerals into a practice that had previously depended solely on the use of plants, he effectively invented chemistry, and then combined it with medicine. He was a brilliant diagnostician, developing treatments for goitres, elephantiasis and syphilis, which at that time was at pandemic proportions. Writing about the complaints of the miners he visited, and treating their silicosis, he became the first person to work with industrial diseases, while for his understanding of the causes of sepsis, and for anticipating the theory of germs, he has become known as the 'father of toxicology'. He was similarly far ahead of his time in his holistic approach, treating people in the round and teaching the contribution of mental wellbeing, and of moral conscience, to overall human health.

Paracelsus ascribed to a complex cosmology, some of which might not be out of place in modern physics. 'Heaven is man', he wrote, 'and man is heaven, and all men together are the one heaven'. The universe, in his conception, was bisexual and homogeneous, time was cyclical and 'above' and 'below' were substantially the same. All creation proceeded from decay, and life permeated all matter. This principle extended to the stars and the jelly associated with meteorites was seen as their excrement, thrown off in a process of purifying themselves. Paracelsus devised his own name for it, combining two words for 'nostril', the Old English *nosthyrll* and the German *nasenloch,* to make *nostoch*. It was, literally, star snot.

It was not long before the word was being widely used as an alternative to star jelly, but if the substance was not meteoric in origin, what else might it be? One of *The Gentleman's Magazine* correspondents was clear: 'The substance is called Nostoch', wrote 'DT' in April 1776, 'and has been ranked amongst plants . . .' This botanical school of thinking ranged over a number of possibilities but, with close observation of the places and conditions in which it occurred, came gradually to focus on an alga. Within another seventy years the link had become official. In *Phycologia Generalis*, a 460-page treatment of the group, the pharmacist and botanist Friedrich Traugott Kutzing described twenty-six members of a genus which he named *Nostoc*. The Paracelsian word had completed its transformation from metaphysics to science.

Today, with the multiplication of taxonomic 'domains' and 'kingdoms', we would no longer consider *Nostoc* to be an alga, not even to be 'ranked among the plants'. It belongs instead to a group known as the cyanobacteria, or more commonly, and somewhat misleadingly, the 'blue-green algae'. They are one of the earliest, and simplest, forms of life on earth; organisms whose single cells contain no nucleus or any other form of membrane-covered organelle. For countless millions of years they floated in masses on the surface of warm, soupy seas, with nothing much happening in evolutionary, or any other, terms. It was only with the appearance of cellular organisms with a nucleus that evolution took one of its great leaps forward. The mechanism by which this happened is still the subject of considerable debate. Inert though they may seem, the cyanobacteria were, however, helping create the conditions in which later multicellular life could flourish, pushing up oxygen levels in the atmosphere and contributing, some 2.45 billion years ago, to what is now known as the 'Great Oxygenation Event'.

There are currently 2,698 known species of cyanobacteria, though estimates put those yet to be described as at least another

3,500. *Nostoc commune* is one of the commoner species among them, existing for much of the time as blackened, brittle, nondescript crusts in the soil. Though single-celled it joins together in chains which under the microscope appear as curling strings of beautiful transparent pearls. In their desiccated form they can survive for up to sixty years but when damp conditions return the chains of cells coalesce, swelling into a huge gelatinous mass. It is a process that can happen with great rapidity, the substance literally bubbling up out of the soil. 'One of its most remarkable properties', wrote a *Gentleman's Magazine* correspondent, 'is its very sudden appearance. In some places where there was none of it to be seen an hour before, you may find the ground covered with it, if in the mean time a heavy thunder shower has fallen'. Its disappearance, as conditions change, can be almost as rapid.

The colour of these jellied masses can vary; from olive to a dark blue-green, from tawny to chestnut brown, sometimes even purple or bluish-black. The surface can be wrinkled or rubbery or have a bubbly texture and they sometimes give the impression of patches of washed-up seaweed abandoned by a retreating tide. If the startled queries appearing in online gardening forums are anything to go by, it is becoming commoner, cropping up regularly on gravel paths and lawns. With its particular predilection for heavy rainfall following a dry period, this increase may well be a function of climate change as we veer more often from extreme to extreme, with heavy rainfall often following periods of drought.

Could it be, then, that the long-debated explanation for star jelly takes the form of a primitive organism persisting from prehistory; a cyanobacterium from the genus *Nostoc*? It certainly fits well with many of the accounts in colour and texture, as it does in one of the details that has received little attention. Star jelly, it was often reported, gave off a strong odour, variously described as 'unpleasant', 'putrid', 'pungent' or even 'nauseous'.

The Nostoc cyanobacterium also produces a smell, one which in the textbooks is labelled more politely as 'earthy'. It is manufactured by a substance with the ungainly name of '*trans*-1, 10-dimethyl-*trans*-9-decalol', more concisely known as 'geosmin'. The discovery of a scientific explanation for the smell of *Nostoc* also suggests it might be worth revisiting another of the details in the accounts; the stories of it 'glowing'. 'A mucilaginous substance is often found on the damp ground', wrote Robert Hunt in 1881, 'near the granite quarries of Penryn, this is often very phosphorescent at night. The country people regard this as the substance of shooting stars'. The many similar accounts have regularly been disregarded as the distortions of the 'vulgar' but it would surely be worth some closer consideration of these claims before dismissing them out of hand. Phosphorescence, or what we would now call bioluminescence, is, in fact, a well-known feature of some marine cyanobacteria. Could it also appear in terrestrial forms? *Nostoc punctiforme* is a rather specialised member of the genus living entirely inside a fungus called *Geosiphon pyriformis,* with which it has a symbiotic relationship. Under certain environmental conditions the cyanobacterium emits light measured at a wavelength of up to 674 nanometres. This puts it in the infra-red spectrum and therefore invisible to the human eye, but it does raise the question of what light-emitting properties might exist undiscovered in other species, only appearing perhaps under particular circumstances or at different developmental stages. There is sense in the suggestion of Paracelsus that we should listen to and learn from 'folk tales'.

Glowing or not, the evidence seems clear that *Nostoc* accounts for many of the occurrences of star jelly. It seems equally certain that it does not cover them all. Star jelly has been described from very different conditions from those in which *Nostoc* can flourish and in a variety of colours and forms. For every account of a green or brown or purple star jelly, there is another in which it is colourless and transparent, or white and completely opaque. It was this last form that was taken by the jelly on the one occasion when I was lucky enough to encounter it myself.

* * *

I come from a large extended family, set in an area of south London where such tribal affinities were once important. The old grandmaternal matriarchy is long since gone, and the clan is geographically scattered, but in other ways we remain close and it is increasingly a younger generation that has kept us together. Once a year, until Covid interrupted everything, we gathered for 'Gilfest', a celebration whose name overlooks the fact that few of us still bear the name of 'Gilbert'. In 2012 our encampment of campervans, tents and trailers was in a field at Britchcombe Farm in Oxfordshire. It was only a month or two since the successful London Olympics, so we had chosen a 'mock Olympics' as our theme, with each of our constituent parties contributing an event or two. Thus it was that over a raucous couple of days we competed at welly throwing, the slow sprint, human dressage, synchronised sitting and various other under-recognised field and track events.

The setting was a striking one. The Vale of the White Horse stretched away to our north through neatly farmed fields, pierced here and there by picturesque villages, until dropping down to the banks of the infant Thames. Immediately in front of us, by contrast, rose the sudden scarp of the Lambourn Downs. Behind the farmhouse, which inserted itself into a fold in these hills, the slopes were darkened with dense woods, but just to the east of this they were open, chalky and floriferous. An outlier known as Dragon Hill is one of the many places where Merlin is said to still lie asleep and behind it, on the main downland flank, is the famous Uffington White Horse, some 374 feet long and two or three thousand years old. Looked at from close quarters it is more of a series of disconnected gashes in the turf, but from a distance it takes on a flowing equine form, as though someone had been shouting instructions through a primitive megaphone from a mile or so away. It is a remarkable composition which as well as being ancient, seems strikingly modern.

Along the top of the downs runs the Ridgeway, a 5,000-year-old track sometimes described as Britain's oldest road. One afternoon we took a break from our pseudo-athletic exertions to walk along it, reaching as far as the Neolithic long barrow known as Wayland's Smithy, before descending to walk back through the Vale. It was towards the end of this excursion that I found the star jelly. It lay on the ground beside the path, close to a little strip of boundary wood in almost the last field that we crossed. It was an opaque and rather unappealing lump, pure white and more like a blancmange than a jelly. It quivered as I scooped it up and there was enough of it to fill the palm of my hand. I carried it with me back to the campervan, planning to save it until we got home where I could submit it to some unspecified but vaguely scientific investigation. Lacking any obvious container, I washed out an empty baked bean can and stored it in that. It seems, looking back, a little haphazard, but I am comforted by an account I came across in the June 1910 edition of the journal *Nature*, in which T. McKenny wrote:

> In 1908 I was with my wife and one of my boys on Ingleborough where we found the 'pwdre ser' lying on the short grass, close to the stream a little way above Gaping Ghyl Hole. For the first time I felt grateful to the inconsiderate tourist who left broken bottles about, for I was able to pack the jelly in the bottom of one, tie a cover on, and carry it down from the fell.

A kindred spirit, I felt, and another example of 'needs must'. In my case, however, there were consequences. Our van is small and when the double bed is unfolded it occupies the available space entirely. For the ensuing night I stowed the can and its precious cargo outside and behind a front wheel. By morning it was gone; not just the star jelly but the can as well.

The gathered Gilbert family is a highly organised assemblage, and everyone in it likes to be useful. So useful, in fact, that if they haven't been ascribed a specific role they will take the trouble to invent one. At a frankly unreasonable hour in the early morning, one member had decided to clean up the encampment. I do have a name but it would be unkind to reveal it. Spotting an empty baked bean can behind the wheel of a van, which implies to me a very close level of inspection, they added it to their black plastic rubbish bag, a bag that was to join many identical bags thrown into the farmyard skip. I did spend some subsequent time rooting through them but it was an unsuccessful and ultimately disheartening search. I had waited sixty years to come across the star jelly, and I have never seen it since.

* * *

Had I managed to take a sample of star jelly home with me, I am not, in truth, at all sure what I might have learnt from it. I had already discounted *Nostoc* as a full and final explanation for the substance and my attention was now turning to something else in the pseudo-botanical field. Referenced repeatedly across the years, and still cited today, was another enigmatic group of organisms known as the Mycetozoa, or, more prosaically, the slime moulds. If *Nostoc* and the cyanobacteria demonstrate a degree of taxonomic uncertainty, then the slime moulds become a full-blown identity crisis. For part of their life they take the form of a rather unprepossessing goo, and were therefore long considered an aberrant fungus. They also, however, have the disconcerting characteristic that they can move. Advancing at a speed of 1.35 millimetres a second, the fastest known rate for any micro-organism, their progress could, in relative terms, be described as rapid. Today, in what is almost certainly not the last word on the matter, they are considered to be 'amoebozoans', lumped into the mixed bag kingdom of the Protista. In other words they have, as Amy Jane Beer put it in an article on the subject, 'no fixed address in the tree of life'. Proposed as the solution to one mystery, they seem to be creating another.

The slime moulds are simple single-celled organisms which, as part of their life cycle, merge into a single multi-nucleate mass. Thousands of them conglomerate to become, in effect, a single vast cell. With its enhanced ability to share both nutrients and information, this amoeba-like 'plasmodium' grows with great rapidity, reaching, in some species, more than a square metre in size. Locating food through airborne chemical signals, the whole porridge-like mass flows towards it, engulfing whatever micro-organisms lie in its way; bacteria, yeasts, other protozoans and even fungi. It is, more than anything else, the perfect micro-model for the 'Blob'.

The fanciful but rather unfortunate popular names applied to some of the more familiar species give a fair idea of their appearance: the 'tapioca slime mould', the 'scrambled egg slime mould', the 'dog vomit slime mould', and, not to be confused with it, the 'dog sick slime mould'. They can appear in the soil, in lawns and grasslands, on woodland floors and, commonly, on rotting wood. It is here that I have encountered the sulphur yellow *Fuligo septica*, topping a damp and decomposing stump and dripping from its sides. It would be almost beautiful were it not for that unpleasant hint of some sort of bodily excrescence. Walking the Isle of Wight coast path with friends more recently, I had a notable encounter with *Mucilago crustacea*, the dog sick slime mould, and perhaps the commonest of our native species. We had rounded the easternmost tip of the island and were now climbing Culver Down, a sudden and striking obelisk-topped prominence that ends in plunging chalk cliffs and the open sea. It was a wet autumn day, with that needle-thin rain that seems determined to penetrate even the most supposedly waterproof of outer garments. A new squall hit us as we reached the top and began our descent over soaked turf slopes towards Sandown. Almost doubled over as I was, I was ideally placed to come face to face with startlingly large lumps of a shapeless white mush that was dotted across the downs. For the most part opaque, they were topped, or sometimes edged, with a transparent runny mucus like the white of an undercooked egg. This, of all the Mycetozoa, I thought, was the one most likely to be taken for star jelly or *pwdre ser*, the snot of the stars.

There was a strangely serendipitous occurrence while I was writing this chapter, a moment of almost Jungian synchronicity. As a Quaker living in east London I am a member of the Wanstead Meeting. Its fine Meeting House has large windows overlooking a simple burial ground that is surrounded by one of the southern fragments of Epping Forest. Watching those fringing trees, solemn in summer, restless and fidgety for much of the rest of the year, has sustained me through many a silent hour. Dealing with some correspondence relating to these grounds, I came across the name of Gulielma Lister – botanist, former member of the Meeting and, as I was to learn, the world authority on the Mycetozoa of her day.

* * *

Born at Sycamore House in 1860, she came from one of the wealthy Quaker families in Wanstead which seemed to characterise that area at the time. Gulielma was not then an uncommon name amongst Quakers, being derived from Gulielma Springett, the first wife of William Penn, founder of the Quaker colony of Pennsylvania. Her uncle was the surgeon Joseph Lister, famous for his pioneering work in antiseptic surgery and it seemed a nice touch that he was a man implementing the ideas first put forward by Paracelsus. Her father, Arthur Lister, was a wine merchant but also a significant amateur naturalist and, with him, Gulielma went on many collecting trips, both locally and from the summer home in Lyme Regis that Arthur and

Joseph had bought between them. They were excursions which her father had christened 'going a-ruggling' and close observation was their keynote as they compiled notebooks, made copious sketches and brought home specimens for microscopic study. Although their interests were broad, Arthur was a particular expert on the Mycetozoa, and it was with him, both in the laboratory and the field, that Gulielma began her work on the group. Together they produced the standard work on the subject, a *Monograph of the Mycetozoa*, detailing all the specimens in the British Museum's herbarium and going on to study the class as a whole. Gulielma was an excellent artist with a fine eye for detail and she illustrated the book with 223 plates and fifty-six woodcuts, initially in black and white but later in full colour. By the time her father died in 1908 she had become the leading authority on the group, adding new species to further revisions of the Monograph and serving as Honorary Curator of Mycetozoa at the museum. She studied collections in both Paris and Strasbourg and even learnt Polish so that she could read the authoritative work that had just been published there. Among her many correspondents was the Emperor of Japan, though as she pointed out, he was too holy to do the actual writing himself. He was another enthusiast of Mycetozoa, and when she named a couple of new species in his honour he responded by sending her a pair of beautiful, enamelled vases.

In more recent times Gulielma has been dubbed 'the Queen of Slime Moulds', a title, I suspect, that neither she nor the slime moulds would welcome. Just as important as her work in this field was her championing of women in science. She was one of the first group of women to be admitted to membership of the Linnaean Society, at whose meetings she would remove her hat, contrary to the convention of the time, to indicate that men and women were equal in science. She was later to become a Vice-President of the Society, as well as first woman President of both the British Mycological Society and the Essex Field Club. She was also instrumental in establishing the Botanical Research Fund to support scientific opportunities for women. I found a photograph of her in her older years; straight-backed, her hair tied in a tight bun. Her expression seemed to combine an unfussy firmness of purpose with an underlying gentleness. There is a half-smile playing around her lips; it has the suggestion of someone who is, very quietly, happy.

Gulielma lived her whole life in Sycamore House and died there in 1949. There is a special process in Quakerism, when a notable Member dies, of producing a 'testimony'. It is not lightly undertaken but has to be agreed and approved through the various levels of Quaker business meeting. The Clerk of the Wanstead Meeting found for me the Testimony that had been written for Gulielma. She had attended the Meeting throughout her life, though it was a different building then, and the Testimony describes her 'coming Sunday by Sunday to take her accustomed place looking out upon the garden . . . even at the age of 88 she would walk there in all weathers . . . and though she seldom, if ever, took a vocal part in the Meetings, her very presence there was a benediction.' Her scientific achievements are then detailed, including the fact that she was 'the greatest living authority on the Mycetozoa, a group of small organisms between the animal and vegetable kingdoms.' Those who knew her best, it concludes, 'will remember her less for her scientific achievements than for her quick wit, her clear judgement, her charm, her half-boyish gaiety and enjoyment of nature and of her friends . . . For sheer goodness, humility and human kindness, it would be hard to find her equal.' I later found her gravestone in the burial ground, just a few yards from that of her father. As is the Quaker fashion, there was no commentary or quote, just a name and dates, but there was a single daffodil blooming in the ground in front of it.

* * *

Though Arthur, Gulielma and the Emperor of Japan were all devotees, the slime moulds remained a comparatively overlooked group until recent years when they shot to something of a celebrity status. These brainless, nerveless blobs, neither animal, fungus nor plant, have a surprising ability, it was realised, for problem solving and for simple learning. They can rapidly solve a maze, finding the fastest and most efficient way through one to reach a food source, usually oatmeal, positioned at the opposite end. They can also be 'trained' to grow through noxious substances, such as salt, caffeine and quinine, acquiring, along the way, a habituation that they will then remember and pass on to untrained moulds with which they are merged. These abilities, particularly that of creating the most efficient protoplasmic connection between multiple points, have led to many scientific applications. Slime moulds have been used in designing road systems and distribution networks, in plotting resource usage, in economic problem solving and even in international border policy. Perhaps only slightly tongue-in-cheek, the Hampshire College in Massachusetts has even established a slime mould problem-solving think tank. But perhaps the most futuristic application has been that of mapping the cosmic web. Using a digital simulation of the 37,000 known galaxies, scientists have used Mycetozoa to map the largely invisible strands of matter that astrophysicists believe to bind the universe, confirming the work with results from the Hubble telescope and thereby reproducing biologically what the cosmos does through gravity. Even the slime mould, it seems, can look to the stars.

The existence of these capabilities in such a simple organism raises fundamental issues in our understanding of intelligence. The slime mould does not solve a maze by trial and error, nor by logical deduction, but by a different process entirely, growing through every possible route simultaneously then withdrawing and consolidating as soon as the most efficient is identified. It demonstrates nonetheless, an ability to respond flexibly to an environment, to solve problems, to make decisions between alternative courses of action and to process complex information. The fact that we have come to regard the possession of a brain as the sole source of intelligence is a consequence of our hierarchical conception of the universe. It is possible that the multi-nucleate functioning of a flowing plasmodium is a prototype for one of the processes incorporated in the workings of the brain. But it is also possible that the whole brain-centric view of intelligence is out-dated and that we should be giving the word a much wider and more open meaning.

* * *

The slime mould was clearly another candidate for some of the appearances of star jelly, but equally clearly it did not explain them all, and other theories were still forthcoming. From 2012 to 2017, the BBC ran five series of a programme called *Nature's Weirdest Events*. The introduction to each episode, filmed through what seemed like filters of brown, showed presenter Chris Packham in a setting of stuffed birds, mounted skeletons and specimen jars, a sort of archetypal mad zoologist's study. The programmes dealt with a wide range of strange natural phenomena, including exploding frogs, self-electrocuting ants, cocoon-covered cars and aberrant weather; though, as critics pointed out, there might have been some barrel-scraping going on by the time they got round to the swimming pigs of the Bahamas in series four. In January 2015 they featured star jelly. Starting from its appearance in a Somerset nature reserve, an event that had been loosely linked to dramatic meteor sightings in the preceding weeks, they dashed through a number of theories before taking a specimen to Dr David Bliss at the London Natural History Museum. It was to provide a different explanation for star jelly.

There had been, to my knowledge, several previous attempts at DNA analysis, most of them proving inconclusive. An effort by the National Geographic Society failed to locate any DNA at all and concluded that the substance was 99% water. A previous treatment of star jelly, on Radio Scotland's *Out of Doors* programme, had led to samples being taken to the Royal Botanic Garden in Edinburgh, where the DNA was found to be contaminated by fungal and bacterial growth. With Dr Bliss, and the sequencing machine at the Natural History Museum, the results were more successful, and located DNA which matched that of a female frog. The substance, it was surmised, was glycoprotein. This is stored in the body of a female amphibian until released at mating when, in contact with water, it swells and becomes the gelatinous mass most familiar in frogspawn. The analysis, however, revealed the presence of a second DNA and it was that of a magpie. The bird, it seemed to suggest, had attacked the frog and either ripped it open or eaten it whole, later regurgitating the unfertilised glycoprotein. Absorbing moisture from the damp ground, it had swollen to form star jelly. While the results were presented as something of a breakthrough, what the programme didn't mention was that experiments without the aid of DNA sequencing had produced a similar result some 360 years earlier.

At a meeting of the Royal Society in the 1660s, Dr Christopher Merrett, a friend of Robert Boyle's, had brandished a specimen of star jelly, or 'star-shoot', as he called it. Some, in the discussion that ensued, had considered it to be the 'mucilaginous matter of a fungus'. Others, arguing that it was found in places where fungi were less common, thought it might be 'some spermatic matter of rams copulating with ewes'. Another suggestion from what might be termed the zoological wing of the argument, was of 'frogs dissolved, especially since bones were sometimes found in it'. As for Dr Merrett, his plan was to obtain more specimens so that he could conduct experiments to discover whether it might be 'an animal or vegetable substance'. In this he was apparently successful, for he demonstrated to a subsequent meeting that the substance was from a frog, which, he said, had been consumed by a group of crows.

The 'regurgitated animal' theory appears regularly throughout the story of star jelly, though the avian suspects attached to it vary. The crow family, of which the magpie is a member, is regularly in the frame, as in the account by another *Gentleman's Magazine* correspondent, J. Platt. Writing in 1776 he begins with another of those confessions of youthful ignorance, having believed until the age of 24 that the jelly-like substance he had found in the fields was the 'dross of meteors'. The fact that his enlightenment happened while he was riding a horse gives it an odd resonance of the original road-to-Damascus conversion. Seeing a group of crows pecking at something, he had turned aside to find it was a frog that they were attacking. 'About this same time', he writes, 'I found in the meadow the bowels of a frog undigested, compact but white as the paper I write upon, though not translucent . . . returning in three days time I saw it changed to that tremulous jelly-like substance . . . This is no fable, Mr. Urban, but real fact.'

John Morton, writing in *The Natural History of Northamptonshire* in 1712, was another naturalist who subjected star jelly to experimentation and came to similar conclusions. Boiling it up, and examining the results, he decided it was the result of an interaction between birds and frogs. 'The above-recited observations', he writes, 'which have been carefully made and are truly related, will by no means admit of either of those hypotheses, that the star gelly descends out of the air, or that it springs from the earth, which others have imagin'd.' In another observation he relates that he had seen a gull, or 'coddymoddy', 'shot down to the ground, that on her Fall . . . disgorg'd a heap of Half-Digested Earthworm, much resembling that Gelly called Star-shot'.

As well as crows and coddymoddies, other birds have earned themselves a place in the line-up over time. A Swedish account named the marsh harrier as the culprit, whilst many respondents to the Radio Scotland programme suggested herons. Writing in *Nature* in 1926, H.A. Baylis described his examination of star jelly collected on Dartmoor in which he had found the oviducts, ovaries and other remnants of a frog or toad. His long list of possible predators included weasels, stoats, badgers, crows and buzzards. The buzzard seems to have become a popular choice in recent decades, perhaps because of its increasing abundance. The 1972 *Collins Guide to Animal Tracks, Trails and Signs* includes a photograph of the jelly which it identifies, very specifically, as 'frog oviducts discarded by buzzards'. Following a rash of sightings in Scotland, a group of naturalists writing in *The Glasgow Naturalist* in 2014, came to a similar conclusion. 'The most frequent source', they suggested, 'is spawn jelly extruded from frogs (or toads) following predation. Like most occurrences of star jelly . . . the observations were on open and exposed hillsides, suggesting that a bird of prey such as a buzzard could have been responsible.'

Whatever the predator, it is the frog theory of star jelly that has, in recent years, come to predominate, and for some it seems to provide the only solution. In the 1968 *Penguin Dictionary of British and Irish Natural History*, the definition of 'Star Slime or Pwdre Ser' is unequivocal: 'A substance found lying on the ground and formerly associated with shooting stars but now known to be the gelatinous remains of frogs and toads'. The matter, it seems, is cut, if not entirely, dried. Faced with the confusion of reality, such reductionism is a normal human response, an attempt to impose some order on the chaos, but in science, as in religion and politics, the urge to find a sole answer to anything should be resisted. I had started my investigation looking for a single explanation for an enigma and learnt along the way that answers can sometimes be multiple. The only certainty I had about star jelly was that it was not one thing but several; a folk name applied to a loose and varied assemblage of phenomena. Bubbling up in grassland with a variety of colours, a distinctive smell and a tendency to appear and disappear very quickly, *Nostoc* closely equates to many of its occurrences. Equally, the often sizeable blobs of opaque white slime mould, or their slathers of sulphur yellow, explain other appearances, especially on a startling first encounter with this ambivalent group of organisms. Expanded amphibian glycoprotein exposed after predation, probably accounts for a great many more including my own lost specimen, though I never got it home for that further experimentation. It also ties in with the numerous local newspaper accounts of recent years. Coming from the Pentland Hills, the Lake District, the Welsh uplands and the Pennine Moors near Oldham and Huddersfield, they fit well with the buzzard predation hypothesis even if one walker did tell the *Manchester Evening News* that 'it looks like frogspawn but definitely isn't'.

Nostoc, slime mould and regurgitated roe, all these can and have been star jelly, as can certain other suggestions that have cropped up over the years: slug eggs, tree sap, stag semen, deer phlegm, sheep afterbirth and otter spraint. In the course of my own investigations, I added another to the list. I passed on photographs of a completely transparent jelly found on a wetland site to the phycologist Chris Carter, who identified them as the 'palmella stage' of an algae; a stage which occurs when the individual algae lose their mobility and congregate together within a swelling gelatinous sheath, though opinions seem divided as to whether they do this in favourable conditions or when just the opposite applies. These clear tremulous gelatinous sheets, I thought, could also have been taken as star jelly.

The one notable absence from all these explanations, however, is any actual reference to the stars. It seems almost as if we had determinedly overlooked this one detail of so many reports; that the substance appeared from the sky. There have been a few efforts to explain what was clearly viewed as a misapprehension of the unlearned. *The Glasgow Naturalist* article suggests that 'a passing bird' might have 'disgorged some jelly which then appeared to fall from the sky', a suggestion which seems to me as unlikely as the meteorite itself. A few other writers propose that people, searching for the place where a meteorite landed, coincidentally came across star jelly and assumed that the one had caused the other. For the most part, however, the idea of a celestial connection, if it is not ignored, is roundly ridiculed. In fact there is a disturbing tendency throughout the whole scientific treatment of star jelly that consists of disparaging the eyewitness accounts. That early reference of Boyle's to the views of 'the vulgar' is echoed over and over again in the use of words like 'unlearned', 'ignorant', 'rustic', or even in a condescending reference to 'country people', as opposed, one supposes, to the erudite urban, or 'the modern learned', as S.G. Goodrich called them, who 'do not subscribe to stars or meteors'. For both Walter Scott and J. Platt, losing their childish belief in the 'dross of meteors' was like a coming of age, while others adopt an almost evangelical tone in their denouncements. 'Most people that I have conversed with on the subject', wrote 'GM', another correspondent with *The Gentleman's Magazine*, 'are of the opinion that this jelly falls from the stars or out of the higher regions of the air which notion, however absurd, many are credulous enough to believe; and I am glad through the channel of your magazine to lay open such an old and vulgar error'.

There is a striking lack of humility in all these comments and they still produce echoes, if less stridently, today. 'The purpose of this note', wrote the authors of *The Glasgow Naturalist* article, 'is to hopefully dispel some of the myths about the phenomenon . . . It is noteworthy that in this current "information age" fanciful ideas and weird speculation are just as rampant as in the past . . . To the uninformed observer strange jelly masses scattered in the countryside will always seem a little mysterious.' Just as the 'myths' were being dispelled, however, and the 'mysterious' being reduced to the everyday, I came across one further, final theory, and it was one that was to return me to the skies.

* * *

The name of Vladimir Ludovich Bychkov seems to be little known in the West. He is Lead Researcher in the Physics Department at Moscow State University and head of the laboratory at the Moscow Radiotechnical Institute. His profile picture shows an intense looking man, but with a broad smile, standing before the imposing domed and pillared frontage of the university. Alongside his picture is a list of his fields of expertise: 'light scattering, polymeric materials, electromagnetics, atmospheric physics and plasma physics'. More prosaically, he has described himself as interested in 'things that fly near the Earth and drop on people'. Presumably this includes ball lightning, for he is Head of the Russian Commission on Ball Lightning and Vice-President of its international equivalent. As well as several books he has authored 154 papers and articles and I scanned through a list of them noting down, for perverse amusement, some of the more impenetrable titles:

'Investigations of Subcritical Streamer Microwave Discharge in Reverse-Vortex Combustion Chamber.'

And,

'Influence of Kinetics on Transversal Discharge Gas Heating Region Size and Form of Supersonic Gas Flow.'

His interest in star jelly began when he was written to by some villagers who had discovered the substance after watching an object fall to earth. After investigating this and a number of other twentieth-century Russian accounts he published a paper in the *International Journal of Meteorology* in October 2005, and another in 2014, on the subject of what he called 'atmospheric gelatinous meteors'.

In taking up water, he argued, clouds are seeded, especially in their upper layers, with particles of dust. These particles contain varied inorganic matter, such as clay and other minerals. They also contain organic substances, including cyanobacterium. Using particle physics, and with the help of some complex equations, he studied the movement of these particles in the cloud, demonstrating, in particular, that the micron-sized coiled threads of a cyanobacteria could remain present for a considerable period of time. During this time they would feed on other organic particles and propagate, forming interconnected coils. When conditions changed, often during periods of rain and thunder, these now complex structures would begin to fall. As they did so they captured water droplets in the air, reaching the earth as a mixture of cyanobacteria, water and slime. They are 'organic meteors', he suggests, whose rotten smell comes from decaying biological components, and whose colours reflect their constituent cyanobacteria. Since many of the cells of the structure are filled with air, their evaporation, once they reach the earth, is rapid, whilst the living organisms within them enter the soil, thus accelerating their disappearance. There might also be a link with ball lightning and the fact that these falls often happen during thunderstorms, it occurred to me, might even account for the stories of some of them glowing.

* * *

The long-running scientific aversion to any celestial connection for star jelly, despite the testimony of hundreds of witnesses, seems worthy of a study in itself. Perhaps there was a need to break away from anything that seemed archaic or alchemical, or perhaps there was a fear to trespass in that strange aerial realm where so much seemed uncertain. It is notable, after all, that one of the cardinal sins of the witch was her ability to fly. Having begun my journey into star jelly with flying objects, with shooting stars and the magic of meteorites, I was grateful to Mr Bychkov. He had restored star jelly, if not to the stars, at least to the clouds, and restored, too, a little of the belief in the ordinary witness. As well as cyanobacteria, slime mould and frog roe, the star jelly could, it turns out, also be something that has lived and bred in the sky, and has fallen down among us. To paraphrase *The Gentleman's Magazine*, 'This is no fable, Mr Urban.'

4

The Underground Mosquito

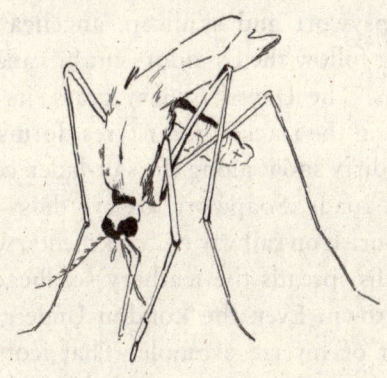

On 10 January 1863 the first London Underground trains began their short and smoky journey from Paddington Station to Farringdon Street. One hundred and fifty years later, the London Tube network had grown to become the largest and most complex in the world. Looked at on the map it resembles a massive web, densely strung at the centre but with threads radiating outwards, through the city and its suburbs and into the countryside beyond. On the Central Line you can travel as far as the Essex market town of Epping and walk from there into ancient forests of hornbeam and beech. In the city's south-west you can ride to Richmond and walk down to the Thames beside Kew or along to the royal deer park and the water meadows of Ham. At the far end of John Betjeman's beloved Metropolitan Line you can step directly off the train into the red kite country of the Chiltern Hills, or turn southwards to follow the water-cress valley of the River Chess. Overall, this far-reaching

network runs for 250 miles, more than half of which, despite its name, isn't 'underground' at all. It is only towards the city that the lines plunge into tunnels, the older ones a shallow cut-and-cover, the newer bored deeper through the London clay and into the chalk beneath.

The shape of our transport networks has always influenced the distribution of our plant and animal species. Waterside plants like gypsywort and skullcap, angelica and hemlock water dropwort follow the corridors cut by canals to colonise deep into cities. The Danish scurvy grass, its seeds carried from the coast in the tracks of car tyres, forms a thick blue-grey haze, like dirty snow, along the salt-laden central reservations of trunk roads. Soapwort, ox-eye daisy and rose-bay willow herb flourish on railway embankments, where the draft of passing trains spreads the feathery seedheads of ragwort from town to town. Even the London Underground system provides a sort of inverse example. That iconic map of its eleven lines depicts only one feature of above-ground London: the course of the Thames in a rather stylised sequence of geometric loops that divide the city north from south. It reveals how the Underground is concentrated to the north of the river, where 241 of London's 270 stations are situated. South of the river it is the overground lines that penetrate deeper into the City's heart. When the London Wildlife Trust began mapping the rapid spread of foxes in the 1980s they found that their penetration of the city reflected these rail routes; underground to the north and above ground to the south. Aided by the abundance of trackside land it was here that they were spreading much faster.

It would be easy to assume, of course, that the truly underground Underground makes no contribution to the city's ecology. Certainly it lacks the richness and diversity of the canal towpaths, roadside verges and railway embankments, but stirring in the tunnels beneath our streets, it does have its own small group of species, and it was one of these, in particular, that I was keen to investigate.

* * *

I have always been drawn to the subterranean world. As a child I dug with my brother a tunnel through the sticky orange clay of our suburban garden, propping up its ceiling with scavenged planks of wood. It went no more than a few yards and filled with water whenever it rained, but to us it seemed a remarkable feat of juvenile engineering. When I left for university, I joined the caving club and explored pot-holes and caverns in the Peak District and the Yorkshire Dales. Years later, living in south Wales, I gate-crashed the Freshers' fair at Swansea University and pretended to be a mature student in order to join their speleological society. When I later confessed my status, they forgave me and we caved together in Fforest Fawr, the Black Mountain and the Brecon Beacons. Since then I have descended working coal mines, disused tin mines and prehistoric flint mines. I have explored catacombs, crypts and the mysterious Cornish chambers known as fougous. Taking my wife on an anniversary weekend to Paris, I punctured the romance by announcing that I had booked us on a tour of the city's sewers. We are, nonetheless, still together.

This world below ground has, for me, both a physical and a psychological compulsion. Nothing else I know of equates to the experience of crawling, hands in silt, boiler suit rubbing the rock, down long, low passageways that suddenly give way to great chambers with fluted walls, rippling calcite curtains, and slowly accreting formations thin as straws or thick as organ pipes. Or of descending the flimsy electron ladders into depths that your headlamp can hardly penetrate, of splashing along stream-smoothed corridors, of feeling the first chill of waist-deep wading or encountering a sump where you must trust the group and the map, submerge yourself and push through the point where the cave ceiling dips below water. When it is part of a running stream or river system like this, a cave can seem not so much the bowels of the earth as its arteries.

The overcoming of adversities, the exertion and the exhilaration, made caving a life-affirming effort, but there was a darker counterpoint. These excursions underground could seem like an intrusion, the entry into a different element or a realm that might be forbidden. It was the place of myth, the Hall of the Mountain King and the Minotaur's labyrinth. It was the location of Lethe and Styx and the place where both Odysseus and Orpheus travelled to speak to the dead. It was, in other words, a flirtation with Thanatos in the darkness in which dreams are formed, and where the subconscious does its continual hidden work. Turn off all the torches when deep in the ground and you encounter a blackness that is absolutely and disorientingly total.

Yet there is life underground. I have never encountered blind cave fish or anaemic amphibians but, exploring the small cave systems of the Gower peninsula, I found roosting colonies of horseshoe bats and became a companion of *Meta menardi*, the cave spider. It is one of the largest British spiders, satin-black with a swollen oval abdomen that can vary in colour from brown to a bright yellow-green. Living in the twilight zone of the entrance passageways, it spins an orb web and hangs its egg sacs on long threads that dangle from the cave ceiling. Having made its acquaintance in these Welsh caves, I was surprised to encounter it later in London, adapting itself to the gothic catacombs of Highgate Cemetery.

Down in the London Underground, the species could be said to include the pigeons. I have seen them, more than once, insouciantly enter a train, travel a stop or two and then step off again, as though it were a pre-planned destination. The brown rat is a more permanent inhabitant, perhaps more so in the past than now, for a photo in the Transport for London archives, dated from 1952, shows rat catchers releasing ferrets in the night-time tunnels; the ferrets a ghostly white against the surrounding darkness. I have a respect, but not an affection, for rats, but the Underground mice are a different matter. They have provided a pleasant distraction on many a late-night journey as they dash along the train pits, their characteristic scuttle interspersed with investigative pauses. They seem unperturbed even when a train passes immediately overhead and more nervous when they risk a sudden rush across a platform seeking some edible piece of litter. Dependent as they are on the crumbs and scraps dropped by passengers, they seem to be always hungry. The Wildlife Photographer of the Year 'People's Choice' award for 2020 went to a brilliant shot by Sam Rowley, who spent five days lying on station platforms until he captured the image of two mice rearing up like boxers in a squabble over crumbs.

Though there is undoubtedly competition over food, the aggressive nature of the mice was somewhat overstated on a notice posted in Farringdon Station in 2012. 'The mice on this station', it proclaimed, 'have been attacking customers. Please place the bottom of your trousers into your socks to avoid being attacked.' The story is still repeated with all seriousness on some websites, seemingly unaware that it turned out to be a joke. What is not recorded is how many people actually obeyed the trouser-tucking injunction. When people were sleeping on station platforms during the wartime blitz, the mice were so numerous that cats were employed to control them. A more recent estimate put their number at 500,000, though the paucity of passengers during the long months of the Covid epidemic, and the consequent reduction in food supplies, will have reduced them significantly. They were, I suspect, always a rather marginal population, living at the edge of their tolerance, and for some time I feared they had gone altogether. Then, as I was writing this chapter, I saw again one of these wonderfully resilient animals, running along a rail in the depths of Bank Station.

Beyond the pigeon, the rats and the mice, there were stories I was hearing of another and more surprising denizen of the earthen urban deep. They came to me first in the form of hints and fragments, second-hand accounts coagulating in the way that an urban myth is formed. They were stories of an insect that had somehow found its way to living an entirely subterranean life. Not like bats or cave spiders in dim entrance passageways or the half-light of cellar or cave mouth, but totally and permanently in the deeply dug tunnels that lay beneath our city streets. Here was a story that was literally unfolding beneath my feet. From that realm so redolent with myth and metaphor, and with its uneasy echoes of the human subconscious, here was a living thing that came to us from out of the darkness. What was its real identity and how had it come to be there? How did it live and feed and breed in such a hostile environment? And was it true, as some had claimed, that there was, in its arrival and survival, a major evolutionary message? How could I possibly resist the mystery of the Underground mosquito?

* * *

Within the great order of the two-winged flies, the mosquitoes constitute a family of their own, with the deceptively soothing name of the Culicidae. The number of its members is a matter of some debate, the well-studied nature of the mosquito giving scientists multiple opportunities to split species, promote or demote subspecies, change scientific names and distinguish the tiniest new details of difference. Consequently, where my older books put the worldwide number of mosquito species at around 1,500, more recent accounts have more than doubled that figure. As well as the more familiar bloodsuckers they include exotic species that steal their food from the mouths of ants or feed only from insects trapped in pitcher plants. For species in the UK, the most authoritative work was written in 1938 at the behest of what was then the Natural History section of the British

Museum. John Marshall's *The British Mosquitoes* began as the revision of another author's earlier work but grew to become a book of its own and remains, some seventy-five years later, the standard text. Despite this, only 1,000 copies were printed and there was no second edition. I was pleased therefore to be able to handle a copy at the British Library, a large, red-bound volume illustrated with photographs, paintings and abundant line drawings and still showing its original price of £1. Marshall details thirty-one species of British mosquito, a number which subsequent work has raised to thirty-four. They occur in marsh and moor, coastal habitats and saline pools, ponds and lakes, woods, gardens and 'wasteland'; and in my bathroom. There, recently, I found *Theobaldia annulata* and though not a connoisseur of the finer points of mosquito identification I knew it from the bold black and white stripes on its legs. It had come into the house to hibernate and, as it is the main source of winter mosquito bites, I afforded it the appropriate respect.

Our human objection to mosquitoes seems to start with their very appearance, for they share with crane flies, harvestmen and house spiders, the feature of long, sprawling legs. It is an attribute to which humans seem to have an innate aversion, linked in the traditional Scottish prayer with all sorts of other nocturnal manifestations:

From ghoulies and ghosties and long-leggedy beasties
And things that go bump in the night
May the good Lord deliver us.

The feature is well-described by D.H. Lawrence in his poem 'The Mosquito', itself a rather long and sprawling affair:

What do you stand on such high legs for?
Why this length of shredded shank
You exaltation?

Is it so that you shall lift your centre of gravity upwards
And weigh no more than air as you alight upon me,
Stand upon me weightless, you phantom?

It may be, as Lawrence suggests, that the long sprawling legs spread its weight and help to make it less detectable when it alights to feed. It is a suggestion I have not seen in any scientific text, but science and art do, after all, arise from the same imagination. The insect, he goes on to say, is a 'ghoul on wings', and the wings are another of its distinguishing features. Long, transparent and thinly membranous, they differ from those of other flies in having their veins and their leading edges thickly clothed with scales. Depending on sex and species, they can beat these wings as many as 600 times a second and in so doing produce their familiar and persistent whine. A sound simultaneously menacing and monotonous, it is described by Lawrence as 'your small, high, hateful bugle in my ear'.

'Why do you do it?' he goes on to ask. 'Surely it is bad policy/ They say you can't help it.' This time the poet is right in only a limited sense, for the whine is something the mosquito is specifically designed for. In 1902 British scientists discovered a special comb-like organ against which the base of the wing rubs, thus producing the sound in the way that children once used to put tissue paper over a comb and blow through it. It would be several more decades before the function of the whine was established. The female mosquito has larger wings which beat more slowly and therefore at a lower pitch. The smaller wings of the males, conversely, beat faster and at a higher pitch. A female can not only detect the male of the species by his pitch, but use it to distinguish him from other species of mosquito. When they locate each other, both will adjust the speed of their wing beats, the female increasing her pitch and the male decreasing his, until they meet somewhere in the middle. It is the love song of the mosquito, and if it matches well enough, they will mate.

The most remarkable part of the mosquito anatomy has to be the elongated mouth parts, its 'proboscis'. In my naivety I had assumed that it took the form of a syringe with a simple blood-sucking process consisting of 'insert, suck and withdraw'; the reality is both more complex and more efficient. There are six parts to the proboscis, the whole assemblage protected within a single retractable sheath. Two of these parts have projecting serrations like a saw, so sharp that they cannot be felt as they go in, cutting into the skin with a quick to and fro motion. Two further organs are then inserted to hold the sides of the wound apart, just as a surgeon might use a retractor in the course of an operation. Only then is the sucking tube inserted, tough but flexible enough to probe about inside the flesh, the chemical receptors on its tip guiding it towards a vein. The sixth and final organ then releases a saliva that prevents clotting, thus reducing feeding time and the likelihood of the insect being detected mid-feast. It is this anticoagulant that produces the later allergic reaction causing the soreness and swelling of a mosquito 'bite'.

'Such gorging', Lawrence called it, 'such obscenity of tres-pass', as he watched the insect drinking. Its belly, swelling and darkening, it will take in as much as three times its own body weight, simultaneously condensing the blood it is ingesting by extracting and excreting its water content. Thus, 'obscenely ecstasied' it might continue indefinitely were it not for the receipt of a specific chemical signal. When one researcher managed to identify and remove it, he watched the mosquito feed until, liter-ally, it burst.

The opponents of evolutionary theory once used to cite the human eye as an organ so complex in structure that it could not, in their view, have evolved as a consequence of multiple random mutations. The mouthparts of a mosquito might prove a less beneficent example of the work of a creationist god, but they are surely just as awe-inspiring in their complex structure, their outstanding efficiency and their exemplification of an amazing evolutionary process. This complex organ belongs only, however, to the female of the species. It is only she who is a blood drinker and while the male, also, can only take liquid foods, he finds them in nectar, fruit juices and sap. His proboscis is consequently much simpler.

For the female it is not just the proboscis that is finely attuned to its task, so too are the features that will guide her towards her prey in the first place. They include the astonishing total of seventy-two receptors in her antennae, interpreting scents and other chemical signals. She can detect CO_2 from 200 feet and has a special attraction for the lactic acid released in sweating. Unfortunately for us, she is attracted just as much by all the things we use to mask our perspiration: deodorants, soaps and perfumes, especially those with a floral scent. It is a well-known and much bemoaned fact that some people seem more attractive to mosquitoes than others. Those particularly on the hit list include beer drinkers, heavy breathers, pregnant women, those with abundant skin bacteria and anyone in the O blood group. The mosquito's use of additional visual cues draws her to particular colours, especially red, orange and black, and once she approaches her potential prey, heat sensors will help her detect the most exposed, and therefore vulnerable, parts of the body.

Such elaborate mechanisms would not have evolved without an important function and this blood feed of the female is vital for the continuation of the species; she needs it and its proteins to complete the process of egg production. In most cases, and the exceptions become important later in this story, the female mosquito needs a blood meal immediately before laying. She will often fly several miles to find one and, once sated, the eggs will ripen in her body and she will seek out a suitable place in which they can be laid. The preference for many species is for some form of shallow, temporary or stagnant water, or even areas such as scrapes or tree boles where water is later likely to gather. Puddles, water butts, old kettles, abandoned buckets, car tyres and neglected bird baths have all served as breeding grounds for the mosquito. Such places are less a pronouncement on its unsavoury character than a mark of the mosquito's survival skills, for they are precisely the sort of inhospitable waters that will be free of fish, the main predators of the mosquito's eggs and larvae.

Laid on the water's surface, the eggs are coated with air bubbles to keep them afloat. Often they form a small, loose assemblage, but in some species are joined together in a floating raft, rather resembling the hull of a boat. The larvae that hatch from them are little more than segmented tubes, with a dark head at one end and a tuft of hair at the other. At this same end is the respiratory tube which they will hold up to the water's surface as they lie immediately below it. Mosquitoes fall into two main groups with names like warring tribes in a Shakespearean drama: the Culicines and the Anophelines. Distinguishing between them has been important in the study of mosquitoes, for it is the Anophelines that carry malaria, and one easy way to tell the two groups apart is through the posture their larvae adopt in the water. The Anophelines lie parallel to the surface, the Culicines hang upside down from it, as though standing on their heads, with the end of their body and its bristle of hairs just breaking the surface. These two different postures reveal two different feeding habits. The Anophelines, whose larvae have the remarkable ability to swivel their necks to 180°, feed only on surface matter. The Culicines randomly sift floating particles for food but can also nibble at algae, dead leaves or abandoned larval skins. John Marshall, with a Georgian gentleman's attempt at wit, described it as the difference between 'dining table d'hôte and dining à la carte'.

There was, at the back of the garage in my parents' garden, a water butt that held a particular fascination for my brother and me. It was busy with what I now know were Culicine mosquito larvae but which we were then perfectly happy to know as 'wrigglers'. Poke the water and they would make their way to the bottom, en masse, in a series of jerky contortions propelling them downwards. Remaining there for a while they would then gradually drift back surface-ward where we could begin the game all over again. I suspect we were proud of our disruptive power. Among these larvae there were always a number of pupae, resembling a comma but with an over-sized dot. At close quarters this fat curled pupa can be seen to pull its way out of its old larval self, leaving only an empty skin behind, in a manner which could have been the inspiration for the *Alien* franchise. This new embodiment of the mosquito is unable to feed and rests at the surface growing darker as the next transition begins within. After about a week it will reach its fulfilment. Just above the water line, the adult mosquito will appear from out of the head end of the pupa. It looks less like an emergence than an expulsion, the adult insect being forced out in a series of contractions as though it were being vomited. At least three times larger than the tight package in which it has been contained, it seems remarkable that it could ever have fitted, especially as those gangly legs unfurl and extend from its sides.

For the adult male mosquito, life is not necessarily nasty and brutish, but it is certainly short. It consists almost entirely of food and sex. To increase the chances of the latter, the males form up in columnar, dancing clouds above a prominent feature such as a chimney, a tree or even a person. These swarms certainly make the males more visible and, I suppose, it is possible that their collective song will carry further. These mosquito 'clouds' have been known to extend as much as a thousand feet into the air and from a distance can resemble rising smoke. In 1807 just such a cloud was seen ascending from the steeple of a church in Neubrandenberg, Germany. Since the church was, at that time, being used as a powder magazine, the inhabitants fled the city until an enterprising local scientist provided the entomological explanation. Explosive or not, the swarms function to attract attention, and the female flies towards them to participate in the mating dance. Once she has been fertilised, the male's life work is done and he will die within a few days. The female, on the other hand, will live for several weeks or months, or even across an entire winter, with some species like the *Theobaldia* entering a house to hibernate. The sperm from one mating will be sufficient for numerous batches of eggs, for each of which the female will need another blood feed, and it is in this moving from feed to feed that, from a human perspective, the problems really begin.

* * *

I have thus far, and with only a few lapses, attempted to describe the life and times of the mosquito with some objectivity; its wonderful complexity demanding, I felt, a measure of awe and respect. There comes a time, however, when my genetic affiliations must have their sway. As far as the human species is concerned, this is the creature that has been described as 'the most dangerous animal on Earth'. Even scientists seem to lose their objectivity in the face of it, outdoing each other in the florid names they have bestowed upon its various species: *Culex molestus*, *Culex perfidiosus*, *Psoropha horridus*, *Aedes excrusius* and *Aedes irritans*. The mosquito bite, however, is far more than an irritant. From blood feed to blood feed, it is the vector for numerous diseases, their names forming a long and disturbing medical litany: yellow fever, dengue fever, West Nile fever, Lassa fever, encephalitis, elephantiasis, zika virus, mayaro virus, chikungunya. More than any of these, however, it is for its role in spreading malaria that the mosquito has made its unfortunate name.

Malaria has been named the biggest killer in human history, outnumbering, even, the slaughter we inflict on each other. Some accounts hold it responsible for as many as half the deaths in human history; an estimated 52 billion over the course of 200,000 years. It is a claim that has been hotly debated with other experts putting the figure closer to 5% or 10%. These lower figures still represent a staggering impact for a single disease, and one which once affected not just many more people but a much larger portion of the globe. Even at the turn of the century, deaths from malaria were still running at a million a year, while the 2020 World Malaria Report puts the current annual death toll at 400,000. It is a burden that falls heaviest on the poorest countries, and on the poorest people in the poorest countries, and particularly on the young. Two thirds of current malarial deaths occur among children under five.

While assessing the impact of the malarial mosquito, we should also consider how we have offered it a helping hand. Our most important move, from the insect's point of view, was that from mobile hunter gatherers, a way of life we had followed for several hundred thousand years, to settled farmers, a mere 12,000 years ago. With this came an increasing manipulation of local environments, deforestation, land clearance, the introduction of irrigation, the diversion of waterways and an increasing population density, further accelerated by the growth of cities. All of it was a favour to the mosquito and the malarial plasmodium. As historian and writer Timothy Winegard put it: 'While agriculture led to a bounty of advancements across human socio-cultural systems, including the written word, it also tampered with and unleashed nature's biological weapon of mass destruction – the mosquito. Cultivation was shackled to a corpse.'

It is a rather Gothic pronouncement but one that serves as a reminder that all our interventions in the natural world have ecological consequences. Further assistance to the malarial cause has arisen from our appetite for campaigns, conquests and colonialism. In his 485-page book, *The Mosquito: A History of Man's Deadliest Predator*, Winegard sets out what he sees as the insect's impact on history, much of it arising from such militaristic endeavour. Malaria, he argues, hastened the demise of the great Mesopotamian civilisations of the Babylonians, the Hittites and the Assyrians. In 492 BC, when Darius was leading the great Persian army towards Athens, it was a combination of dysentery, typhoid and malaria that halted his advance. When Xerxes attempted the same feat just twelve years later it was to result in the loss of 40% of his army to malaria, and to a swing in the balance of world power towards Greece. Having helped to make a power of Greece, the mosquito also helped to break it. When the Athenians invaded Syracuse in 413 BC with 40,000 troops, they were deliberately lured into malarial swamps, where the disease killed or incapacitated 70% of them. It would have been a terrible end, this mass dying in the marshes, and an early use of the biological warfare which we wrongly think of as a peculiarly modern phenomena. When Roman power replaced that of the Greeks, they used a similar tactic, deliberately drawing invaders into the mosquito-infested Pontine Marshes surrounding Rome. It would be more than a millennium before the link between malaria and the mosquito was scientifically established and it was the view then, and for long after, that the disease was caused by foul vapours or 'miasma' arising from such wet and boggy places with their mists and marsh gases and the smell of rotting vegetation. From this belief came the name 'malaria', formed from the Italian words for 'bad air'. In response to it, the wealthier Romans had built their houses upon the city's higher ground, a practice that would come to be emulated across the empire. Even up to recent times it was possible, in London, to map how affluence increased with altitude, a pattern that was only interrupted when the de-industrialised waterside became fashionable.

Despite their precautions, the Romans suffered. The emperors Octavian, Titus and Julius Caesar, all experienced malarial bouts. Vespasian, Titus and Hadrian died from it while the whole of the late Roman Empire, Winegard argues, was fatally weakened by it. It was, he goes on to claim, malaria that halted the Mughal advance, that led to the decline of the Khmer civilisation and that contributed to the near eradication of the indigenous populations of South America and the Caribbean with the arrival of the European colonists and their imported African slaves. Further north, malaria assisted in the foundation of the United States. The long-term settlers, being more inured to the disease, gained an advantage in the War of Independence where it 'consumed the British forces and ultimately decided the fate of the revolution and, by extension, the world as we know it.'

Winegard's enthusiasm for his thesis leads him, perhaps, to overstate his case, something which is always likely to happen when you view the world through a single lens. Nonetheless, it is hard to escape his conclusion that the mosquito has been a major player in human history and, both socially and politically, has helped to shape the world in which we live.

* * *

The first mention of mosquitoes in the London Underground dates from the wartime bombing of the early 1940s. The Tube stations had first been used as shelters during the Zeppelin raids of the First World War, the government promoting their use with the slogan, 'Bomb proof down below'. During the early months of the Second World War the attitude was very different. Fearing what they called a 'deep shelter mentality', the government banned the overnight use of tube stations believing it would be bad for morale, would spread disease and panic, and that, once down there, the population might simply refuse to come back up again. John Colville, Churchill's private secretary, wrote in his diary that although the Prime Minister was taking refuge in a disused Underground station himself, he was 'thinking on authoritarian lines about shelters and talked about forcibly preventing people from going into the Underground.' While the government was concentrating on such surrogate measures as the distribution of ear plugs, the people took matters into their own hands, gaining access to the tunnels by the simple device of buying a 1½d travel ticket and then refusing to resurface afterwards. Carrying deck chairs and rugs they would alight at a station and, if it was already overcrowded, take another train to find space at the next. By September 1940, and the beginning of the Blitz, public pressure was increasing, with incidents like that at Liverpool Street where large crowds gathered outside the station in the evening, demanding to be let in. When the government gave way it was an about-turn hailed by the *Morning Star* as 'an important victory for London people against the Government.'

The 'London people' began to shelter in the Tube in large numbers; 120,000 of them by 21 September according to official estimates, a number which had grown to 125,000 by October. By November at least one London council was issuing official shelter reservation tickets, with the condition that shelterers arrived after 6.30 in the evening and left by 7 a.m. Bringing their blankets and bed rolls, they slept shoulder to shoulder, usually in the clothes they had arrived in, across the platforms and up the escalators. Since the tube network continued to run until 10.30 p.m., a two-foot space had to be left at the platform's edge to allow the genuine passengers to pass. Once the last train had run, and the electric current was switched off, the people would sleep on the track bed as well or put children in hammocks slung across it.

Modern accounts of those nights in the London Tube tend to dwell on their bonhomie and the idea of a 'spirit of the Blitz', as though it were all one continuous cockney party. Indeed, there was a measure of this, with people bringing down wind-up gramophones, singing and dancing, and organising concert parties and amateur dramatics. An official docket from the Passenger Transport Board details the expenditure on Christmas entertainment for children. It includes 100,000 paper hats,

11,568 toys, 120 Christmas trees, £18 worth of 'holly and heather sprigs', a £25 contribution to the Salvation Army and five amplifiers, 'a gift of the American Committee for Air Raid relief'.

Much of this rosy view of life underground arises from the effectiveness of Government propaganda and still prevails today. The truth is, however, that despite the best efforts of many, conditions in the Tube shelters were grim. People were packed unpleasantly close together, sleeping on uncomfortable surfaces, spreading coughs and colds, hardly able to even turn in their sleep, and fearful of what might be going on above. There was competition for places and fights over spaces and sanitary conditions were appalling, with just a few Elsan toilets or buckets, shared between hundreds of people. The experience is described in a film at the London Transport Museum by Len Phillips who lived through it as a child. 'The worst part', he says, 'was the smell . . . the smell of bodies and of other unmentionable things as well.' In addition, 'you could hear everything that was going on up the top, you could hear the bombs, you could hear the guns that echoed down the lift shafts. I hated it, to be frank.' It is the smell that figures, too, in the account of Evelyn Rose, also a child at the time. 'The stench was unbearable. The smell was so bad I don't know how people did not die from suffocation. So many bodies and no fresh air coming in. People would go to the tube stations long before it got dark because they wanted to make sure that they reserved their space. There were a lot of arguments amongst people over that.' Nor were the stations and tunnels quite as secure as people wanted to believe and, for some, no shelter at all. At least 316 people were killed by bombs penetrating Underground stations, or by crushes as they rushed to enter them.

The darker side of the experience was captured by some of the official war artists, the most famous among them being Henry Moore. The Tube shelters, he said, were 'places that gave the appearance of death even as they preserved life'. His drawings of long lines of sleepers, side by side in tunnels, wrapped in blankets that look disturbingly like shrouds, are executed in a monochrome grey, sometimes with the addition of a washed-out yellow or red. They show an awkwardness and an unease that is as much about vulnerability as resilience and they have disturbing echoes of the mortuary, the charnel house or the lines of blanketed bodies laid out after an air raid. For me they call to mind the opening lines from Wilfred Owen's First World War poem 'Strange Meeting':

> *It seemed that out of battle I escaped*
> *Down some profound dull tunnel, long since scooped*
> *Through granites which titanic wars had groined.*
> *Yet also there encumbered sleepers groaned,*
> *Too fast in thought or death to be bestirred . . .*

As the war progressed, conditions in the shelters began to improve, both by official effort and by those of the shelterers themselves. The stations were marshalled by wardens, sleeping spaces were regulated, a row of bunks installed and white lines painted to separate the sleeping zone from the passenger gangway. Pumping systems were introduced to deal with the sewage and, with the help of thirty doctors and 200 nurses, first aid posts were established at every station. There were 124 canteens, as well as 'refreshment trains' travelling up and down the lines, which collectively served an estimated 545,454 gallons of tea.

* * *

To the discomforts of shelter living, the mosquito bite was but one more addition. The first complaints were received by the Medical Officer of the Port of London Authority and came from Liverpool Street, the same Underground Station outside which people had gathered to demand overnight admittance. The medical authorities visited and reported catching many mosquitoes, 'some in the act of biting'. Further complaints were soon being received almost daily, ranging from Clapham Common in the south to Hampstead in the north. Investigators established that breeding was taking place in the 'inverts' beneath the platforms, even where the depth of water was only a couple of inches, as well as in deep sumps elsewhere in the system where large numbers of larvae were being found. Further references in official documents detail the authorities' response to the problem. 'Good progress', ran one of them, 'is being made in the delivery of sprays and compressors for aerial disinfection', adding that, 'spraying may be discontinued while the shelterers are sleeping but should recommence during the coughing period in the morning'. The reference to a 'coughing period' creates in just two words a vivid mental picture of life in the shelters in the mornings, as shelterers stir and rise with a grunting and coughing and clearing of throats, gather up their possessions and make ready to face the day.

While the arsenal of chemical controls was being assembled, the staff needed to be kept informed and one way to do this was through the staff newsletter. An article published in February 1941 is written in language that patriotically attempts to mirror the wider struggle being waged in the skies above. 'Paraffin attacks on the mosquitoes' "airfields"', it runs, 'have stopped nuisance raids by this winged enemy of Underground passengers and shelterers.' The Board also needed to keep the wider medical community informed and a letter from 'The Officer in Charge of Tube Shelterers' was sent to 'All Medical Officers of Health' on 8 March 1941. A surviving copy, in the Transport for London archives, is decorated with numerous departmental stamps and scribbled reference numbers, and you can visualise its bureaucratic progress as it is passed from desk to desk. One hastily pencilled note reads, 'Shelter Inspectors to note please', and beside it another adds, 'Noted'. Headed 'Mosquitoes at Tube Stations', the letter is worth reading in full and, once again it seems, 'good progress' is being made:

Measures are being taken to overcome the mosquito trouble at stations. Due to the rise in temperature caused by the presence of shelterers and the opportunity for mosquitoes to obtain a 'blood' meal, the tubes have developed mosquitoes where none existed before. I am glad to say that with the help of Dr. Stock and his mosquito experts, good progress is being made and the nuisance is well under control. A staff of five to ten men has been employed to deal with it and will continue so long as there are signs.

Help will be afforded if, upon hearing of any sudden recurrence at any station, you could arrange immediately to telephone the Board's Building Department. Time is the essence of subduing these marauders and upon receipt of the information the mosquito staff will proceed to the spot at once.

If you could do this, I think that very little trouble is likely to arise in the future.

Among the Medical Officers receiving this letter, some of the older ones may well have given thought to a possible malarial connection. It is now little remembered that malaria was once endemic in England and the collection and publication of annual case numbers had ceased only as recently as 1910. Even this had been followed by outbreaks at the end of the First World War when local mosquitoes were reinfected for a time by the return of sick soldiers. The memory of malaria, as an English disease, was well within the life span of many of those now being bitten in the shelters.

* * *

There are four species of the malarial plasmodium that parasitise the mosquito. The most virulent of them, *Plasmodium falciparum*, requires a temperature of at least 20°C for up to three weeks to complete its cycle, but the milder *Plasmodium vivax* can develop at 15°C and it is this form that carries malaria in cooler climates. Milder it might be, in comparison with the deadly *falciparum*, but its impacts are anything but pleasant. They include recurring paroxysms lasting about eight hours each and consisting of intense bouts of shivering followed by a contrasting high fever and heavy sweats. They are accompanied by anaemia, an intense lassitude and a swollen spleen. In England the disease was known as the ague and the hard feeling of the enlarged spleen as 'ague cake'. It had seasonal peaks in the autumn and spring and children were amongst the most susceptible.

In early accounts, before the link between malaria and the mosquito was established, it is not always easy to distinguish whether it is the ague or another disease that is being described, especially infections that can display similar symptoms, such as typhus, typhoid, influenza or enteric fever. Some of the best clues we have to its identification come from literature and Chaucer, in his *Canterbury Tales*, provides one of the earliest examples. In the relatively brief 'Nun's Priest's Tale', a poor widow has among her few possessions a very fine cockerel, Chanticleer by name, who is particularly proud of both his comb and his crowing. He lords it over seven hens, among whom he has a special affection for the Lady Pertelote. One morning he wakes screaming from a dream. The Lady berates him for his cowardice and, as set out in the translation by Nevill Coghill, expounds her view that, 'Dreams are a vanity, God knows, pure error'. They are the result of nothing more than his ill-health:

> Your face is choleric and shows distension;
> Be careful lest the sun in his ascension
> Should catch you full of humours, hot and many,
> And if he does, my dear, I'll lay a penny
> It means a bout of fever or a breath
> Of tertian ague. You'll have your death.

'Tertian' ague is another name for the malarial condition, describing the alternate day periodicity of *Plasmodium vivax*. Just as interesting as the diagnosis is the Lady Pertelote's prescription:

> Worms for a day or two I'll have to give
> As a digestive, then your laxative.
> Centaury, fumitory, caper-spurge
> And hellebore will make a splendid purge;
> And then there's laurel or the blackthorn berry,
> Ground ivy too that makes our yard so merry.

It is a rather poisonous assemblage of plants, notwithstanding the ground ivy which presumably makes the yard merry not just because of the prettiness of its flowers but because it was once used, in place of hops, in the brewing of beer.

Alcohol was itself another substance regarded as a cure for the ague, something which is reflected in Shakespeare's *The Tempest*. The butler Stephano, one of the courtiers of the King of Naples, is among those shipwrecked on Prospero's island during a storm. As he staggers drunkenly about, singing and clutching a bottle, he encounters Prospero's monster-servant Caliban, who is out collecting wood. Caliban is terrified by the encounter and begins to tremble, leading Stephano to the conclusion that he has the ague and must undergo the cure:

> [He] hath got, as I take it, an ague . . . He's in his fit now, and does not talk after the wisest. He shall taste of my bottle: if he hath never drunk wine afore and it will go near to cure his fit . . . Open your mouth this will shake your shaking . . . if all the wine in my bottle will recover him, I will help his ague.

I have always found *The Tempest*, perhaps along with *King Lear*, to be one of Shakespeare's most powerful plays. Less fortunate students of English might also have been required to study Samuel Butler's *Hudibras*, a long, mock-heroic narrative poem published in 1663. Written in the aftermath of the Civil War, it is a partisan polemic, a satire against Roundheads, Puritans and Presbyterians. Butler, in other words, was a passionate Royalist. Where trembling with fear is, in *The Tempest*, mistaken for a symptom of the ague, here it is love, or more strictly, infatuation, that is conflated with the disease. A high-born lady rejects the amorous advances of a knight, telling him that love is fickle and follows the same course as the paroxysms of malaria, though in a different order. It is, she says,

> *. . . but an ague that's revers't*
> *Whose hot fit takes the patient first*
> *That after burns with cold as much*
> *As Ir'n in Greenland does the touch.*

Given this satiric reference to malaria, it is perhaps ironic that Butler's great hero, the recently restored Charles II, was himself to suffer from it. Unlike Butler, however, the plasmodium was not partisan and his arch nemesis, the Protector Oliver Cromwell, was also a victim. By this time a new and more effective treatment for the disease had been discovered in the form of quinine. Derived from the bark of the cinchona tree, its curative powers had been revealed to Jesuit missionaries by the native peoples of Peru and it became known, for a time, as 'Jesuit powder'. It was not a name that would endear it to the puritan Cromwell. Insisting that he did not want this 'Pope-ish remedy' or to be 'Jesuited to death', he died of malaria instead.

* * *

Another high-born person who was famously afflicted was the adventurer Sir Walter Raleigh. Facing execution in the Tower of London, his final words to the axeman were, 'Let us dispatch. At this hour my ague comes upon me. I would not have my enemies think I quaked from fear, Strike, man, strike!' Notwithstanding these examples among the well-known and the wealthy, however, malaria in the UK was primarily a disease of the poor. It was also a particular part of the country that was most affected. With the exception of just one or two other pockets, Morecambe Bay among them, malaria was almost entirely associated with a 300-mile strip of the Eastern English seaboard, running roughly from Hull to Hastings. Covering eight counties it included the lower Aire and Ouse valleys in East Yorkshire, the low country and Fens of Lincolnshire, the Wash and the Norfolk Broads, large swathes of Cambridgeshire, Essex and the Thames estuary, and down through Kent and East Sussex to Romney Marsh, the Rother, Pett Levels and Pevensey. Here were the low-lying flat-lands with their multiple estuaries and winding creeks, their ditches and dykes, their flooded fields and splashy meadows, puddled and pooled, that the mosquito loved. There are five species of them in this country capable of transmitting *Plasmodium vivax* and among them, breeding in the brackish coastal and estuarine pools, was *Anopheles vivax*, its main carrier.

In an article from July 2018, Greg Bankoff, Professor of Modern History at the University of Hull, labelled this whole area as the 'English Lowlands', a distinctive region characterised not only by its open skies and flat landscapes, criss-crossed with dykes, and dotted with windmills, but also by the common way of life of its people. It was a way born of vicissitude and the travails of living with both the constant risk of flood and storm surge and the effects of recurrent bouts of ague. For these people the 'marsh fever' was part of their culture, a culture which had its own laws and customs and even its own architecture. They were resilient, inventive and highly independent, and with a strong sense of collective responsibility, but to the outsider they were a strange people of pale and sickly complexion, with peculiar habits such as walking on stilts. They were reputed to have webbed feet and repeatedly characterised as stupid, apathetic or fatalistic, a characterisation not unrelated to the perceived effects of endemic malaria. As the anonymous poem, 'The East Anglian Fenman', put it:

> *The moory soil, the wat'ry atmosphere*
> *With damp, unhealthy moisture chills the air*
> *Thick stinking fogs, and noxious vapours fall*
> *Agues and coughs are epidemical*
> *Hence every face presented to our view*
> *Looks of a pallid or a sallow hue.*

With its people weakened by malaria, high mortality was another feature of the region. Though this strain was not itself a killer, it made its victims more susceptible to other conditions and life expectancy was consequently little more than thirty years. The medical historian Mary Dobson has demonstrated from parish records that death rates were two to three times higher in marsh-land parishes than in neighbouring ones. In the pre-nineteenth-century fens of Cambridgeshire and Lincolnshire, malaria was associated with as many as 20% of all deaths while, in estuarine Essex, the annual death rate in marsh parishes could rise to as much as four times the national average with 80% of such parishes recording more burials than baptisms. Visiting Essex in the 1590s, the topographer John Norden found the ague to be rife, 'especiallie nere the sea coastes . . . and other lowe places about the creekes which gave me most cruel quaterne fever'. He was not the only visitor to catch it. Discussing the 'strange decay of the marsh folk', Daniel Defoe reported in his 1722 *Tour Through the Eastern Counties*, that 'gentlemen who went to Essex marshes to shoot wild fowl . . . returned with an Essex ague on their backs which they find a heavier load than the fowls they have shot'.

The 'gentlemen' could at least move on; the marsh folk must remain. They did so because here were some of the best soils in the country, providing work on rich agricultural land with good grazing. But while management of the land and its livestock was conducted by the local 'marshmen', the landowners themselves remained at a safe distance living on higher land. Even the church, it seems, had deserted the locals. According to Mary Dobson there were at least twenty-eight parishes in marshland Essex and North Kent where vicars and curates refused to reside alongside their parishioners complaining, in their returns to the Bishop, that they were 'so violently afflicted with the worst of agues . . . languishing so long under it till our constitutions were almost broke.' Among the most notorious of these parishes, where visitors commented on the 'swollen bellies of the children and their sallow, sickly faces', were those of the Essex Dengie.

The Dengie is a fat, blunt peninsula, set between the River Crouch to the north and the broad estuary of the Blackwater to the south. Occupying about a hundred square miles, it is still described as one of the most remote areas in southern Britain with few villages or homesteads and pierced by just a few thin tracks that terminate in isolated farms. The boundary between sea and land would have been indeterminate and shifting here, until it was defined by the building of the fifteen-foot-high sea wall that runs the fifteen miles from Burnham to Bradwell. It was along this sea wall that I set out to walk with my friend Richard, a companion of many such coastal excursions, and we arrived by train in Burnham, one June morning, along a branch line that seemed a last wistful fragment of the network pre-Beeching. Burnham-on-Crouch is a moderately pleasant village with a meandering high street, some timber-framed inns, and a flocking of yachts down by the quays where their wires whine and their masts tinkle in the estuarine wind. They amass like animals about to migrate, their numbers confirming that this is now a comfortably affluent place. It is hard to envisage Burnham as one of the malarial villages of the Dengie yet, writing in the *British Medical Journal* in 1938, a Dr Wilson could still recall a conversation with an 'ancient' in which he described his sufferings from ague at the age of nine: 'the terrible coldness as he sat over the fire with his teeth chattering, and how his mother used to put rugs round him to try and get him warm. Soon afterwards he would be just as hot . . . It was severe enough to keep him away from school for six months.' In the eighteenth century this had been one of those parishes whose priests had refused to remain in permanent residence, the vicar here insisting to the Bishop that it was 'for the protection of my and my family's health'.

Richard and I left Burnham and its beached or anchored yachts, walking eastwards into starker country along the earthen sea wall. It was a cloudy and blustery day but the banks were nonetheless busy with bees and our passage scattered clusters of small heath butterflies, bursting from the surrounding tall grasses and blowing like a litter of tiny leaves along the path ahead. After five miles, and the negotiation of such indentations as Winkle Bay, we reached Holliwell Point opposite Foulness Island. Here the estuary gave way to the open sea and the wall turned northward, at first along a short stretch of concrete, with a beach of crushed shells beneath, then recovering its familiar earthen form above flanking salt flats. A uniform grey-green, the colour of sage, they stretched off seaward, their monotony relieved only by penetrating channels and the occasional appearance of a leaning, half-rotten post. At the base of the sea wall the flat beds of sea purslane gave way to straggling shrubs of sea blite, while the banks themselves sported the thick glossy leaves of sea beet, the last purple flowers of salsify and swathes of alexanders, their umbrella-shaped heads now black with seed. A cuckoo burst out of a hawthorn bush and a fulmar provided a surprise passer-by above the flats. It seemed to me that all diversity of natural life was concentrated here on the sea's edge, on the margins of things. Inland it was a different picture and bleak with the undifferentiated indifference of industrial-sized fields that were as featureless as they were depressing. It was through these fields that we turned at the end of the day, heading inland, through a drained and half-dead uniformity to the village of Tillingham and the pleasant warmth of its white and weather-boarded pub.

Though the village is set a couple of miles inland, it was to here that the waters dashed during the disastrous floods of 1953, putting most of the Dengie underwater. Once the sea had broken through, that sea wall had served to retain it, delaying its retreat and sustaining the wretched conditions. I was reminded, too, how some commentators had suggested its erection might have aided the mosquito, replacing areas once regularly scoured by tides with a mass of still and stagnant pools and channels. It was from Tillingham that we were to set out again, this time under the grey sky of a dull October day with the flatlands turned to shades of fawn and dun and khaki and the fields bristly with the remains of cut crops, or pricked with the new growth of winter wheat. From the wall we could see the arriving flocks of Brent geese settling on the sea beyond. A peregrine passed rapidly over the salt flats and the sun broke through briefly, producing a play of light and shadow on a group of distant poplars. Despite this, the 'brooding atmosphere' and all such other travelogue-talk, the walking was, in truth, tedious; a long, straight wall set between landscapes that were, on either side, unchanging. And all day long we could see our destination. The village of Bradwell is incongruously flanked, on one side by St Peter's Chapel, one of the oldest religious buildings in the country, on the other by the giant bulk of a decommissioned nuclear power station. It was towards the red roof of the chapel that our steps were continuously set and all day long it seemed to get no closer. It was hard to envisage now that tide-washed patchwork landscape of the malarial years, with its mix of small woods, reed beds, wet meadow and marshlands, haunted by wildfowl and waders. Here had lived the 'lookers', as the marsh people were mysteriously known, with their cottages, their crops and their cattle, in a time before the huge agricultural machines of the undivided field had taken the place of people. But here too had lived the mosquito, in the typical landscape of the 'English Lowlands', breeding in masses in the brackish pools and drainage dykes and afflicting both adult and child alike as they sweated and shivered their way through the annual bouts of the ague. And still, with all the changes, a dull and desultory air hung over the place. It was easy to see why H.G. Wells had chosen it for scenes from *The War of the Worlds*; and how Alfred Hitchcock had been reputedly influenced by it whilst filming *The Birds*.

* * *

Further south, the watery county of Essex reaches its boundary at the Thames. This estuary too, both on its Essex and its Kentish shores, and as far inland as the metropolis itself, was another part of the malarial lowlands. In London, the Lambeth and Westminster marshes were particularly notorious and epidemics had been recorded here as late as the hot summers of 1857 and 1859. According to the military physician and expert on tropical diseases Sir William MacArthur, 'of all the patients treated at St. Thomas's Hospital between 1850 and 1860, one in twenty was suffering from ague and in bad years the average figure was much higher.' Just downriver, on the Kentish side, lies Gravesend, and in its hospital during these same years, reports MacArthur, about 30% of all patients had cases of the ague. No doubt many of them had come from the neighbouring villages of the North Kent marshes and from its Hoo peninsula that begins immediately to the east of the town. This was one of the very last English strongholds of the disease and bears its own melancholic memorial in the little peninsula churchyard of Cooling.

This, in turn, provided what must be one of the best-known of all literary references to the ague.

It was in this 'marsh country, down by the river' that Charles Dickens set the opening chapters of *Great Expectations*, and it was here in the Cooling churchyard, it is now generally supposed, that the orphan Pip had stood among the graves and learnt the fate of his family, finding,

> for certain that this bleak place overgrown with nettles was the churchyard; and that Philip Pirrip, late of this parish, and also Georgiana wife of the above, were dead and buried; and that Alexander, Bartholomew, Abraham, Tobias and Roger, infant children of the aforesaid were also dead and buried; and that the dark flat wilderness beyond the churchyard, intersected with dykes and mounds and gates with scattered cattle feeding on it, was the marshes; and that the low leaden line beyond was the river; and that the savage lair from which the wind was rushing, was the sea.

I had come here myself as a boy, on one of our family explorations of Kentish churchyards and it must have been an appropriately bleak marshland day for I remember it as if it were permanently enwrapped in a shroud of grey sky, and such a gloomy place that I could well understand Pip's description of himself as a 'small bundle of shivers growing afraid of it all and beginning to cry'.

Still in search of the old malarial landscape, I wanted to revisit the place and when I did so the day was not grey at all, but blue and blustery, with troupes of flat-bottomed cumulus clouds processing overhead and a large half-moon hanging in the daytime sky. The village was surrounded by orchards, not, sadly, the apples and pears of the proud traditional varieties but the modern form with a dwarfed and stunted stock. To the north the land subsided gradually into the marshes, to the south it rose through fields to reach a rim of low hills that would once have marked the edge of the wetlands. The church of St James is rough-hewn and mossy, its walls a mix of the local white ragstone with contrasting nodules of black flint. They are flanked by leaning buttresses but they have the appearance of casually leaning against the church rather than helping to actively hold it up. There is a red roof and a square tower, above which the jackdaws were busy, uttering their high twang as they threw themselves here and there into short bursts of flight in the gusting wind. The churchyard is contained by a low, red-capped stone wall, with strands of ivy occasionally adventuring over its top. There was a line of now bare limes along one side, and of half-stripped sycamores along another. Their remaining leaves rustled metallically, their blown ones tumbled and dashed and caught in the uncut grass between graves. Close to the church, as there always should be, was one of our venerable churchyard yews, not huge, but definitely handsome, a female tree with a heavy crop of bright red arils. Its original trunk had almost rotted away leaving mere fragments which struggled to encase a strong, younger bole within. It was one of those remarkable demonstrations of the tree's regenerative power. Clearly, an aerial root had penetrated the hollow of the original tree and was now growing to replace it. With the last sections of the older tree still clinging on here and there, it was like a snake sloughing off the skin in which it no longer fitted.

The church guidebook specifically stated that the 'berries' of the yew were poisonous so I defiantly ate some. It was with this reminder of mortality that I came upon the slab now known as 'Pip's grave'. It is white, rectangular and curly-headed and its inscription illegible. Laid out beside it, in descending size, are ten tiny tombs, the shape of miniature mummy cases, with three more, slightly larger, on the opposite side of the slab. The date of 1774 is just about visible on one of them. Their colour is a pale chalky white, stained yellow with lichen, the pallid colours of the marsh fever from which these children almost certainly died. They are the remnants of just two families, the youngest of them barely a month old; testament to the depredations of the ague, its impacts upon this area, and, in particular, upon the young.

Dickens often walked here from his home in Higham and if it is, as supposed, the site on which he bases his description, he changes it to suit his narrative. The thirteen graves of Cooling become the

> *five little lozenges about a foot and a half long beside their [parents'] grave and sacred to the memory of five little brothers of mine – who gave up trying to get a living exceedingly early in that universal struggle – and I am indebted for a belief I religiously entertained that they had all been born on their backs with their hands in their trousers pockets and had never taken them out on this side of existence.*

It was from behind another of the lichen-coated slabs closer to the church porch that the convict Magwitch had suddenly reared up, threatening to cut Pip's throat, 'a fearful man, all in coarse grey, with a great iron on his leg. A man with no hat, and with broken shoes, and with an old rag tied round his head. A man who had been soaked in water, and smothered in mud, and lamed by stones, and cut by flints, and stung by nettles, and torn by briars . . .' He was clearly displaying the symptoms of the ague as he 'limped, and shivered, and glared, and growled; and whose teeth chattered in his head as he seized me by the chin.' Magwitch had escaped from the 'hulks', the decommissioned and de-masted boats that had been used to house convicts on the nearby river. They were rife with the malaria that afflicted Magwitch, with just one ship, the *Justitia*, recording as many as 157 cases in the first three months of 1847.

I walked round to the front of the church and looked down over the marshes to the estuary where the hulks would have been anchored. Here, unlike the Dengie, the wide expanse remained pastoral, the two miles to the riverside dotted with sheep. On the opposite shore were the structures of industrial Essex, the oil refineries of Shell Haven and Thames Haven and the waterfront at Coryton, lined with huge four-legged cranes whose gibs dipped up and down as if they were a row of giant, metallic giraffes drinking from the river. On this, the Kentish side, things were much more as Dickens might have seen them, the lines of criss-crossing dykes marked with low groups of osier, bramble and blackthorn, the puddled fields bristly with straw-coloured tufts like punk haircuts, and the reeds in beds and channels creating an incessant background whisper as they bent in the continuous wind. Crows fed among the sheep, magpies fussed in the small trees and flocks of lapwing flew overhead uttering their pinging calls as they passed. I had once read that the red dead-nettle was known on these marshes as the 'convict's flower' as it was believed to grow on their unmarked graves. It is not generally a marshland plant and seeing it nowhere about me I puzzled over this local association. Then I remembered that red dead-nettle is one of the colonising plants of disturbed land. It would have grown quickly where the ground was dug and heaped, producing its whorls of pinkish-purple flowers to pick out these anonymous plots from the rest of the surrounding land. It has another country name of 'archangel', so at least their inmates were receiving some sort of a blessing.

* * *

Medical historians have today identified the mosquito of the English lowlands as a species named *Anopheles atroparvus*. When Dr Stock and his wartime 'mosquito experts' identified the specimens they had collected in the wartime tunnels, they were found to be of two quite different species. One was my one-time bathroom companion, *Theobaldia annulata*, the other was *Culex pipiens*. This is a smaller mosquito of rather undistinguished appearance, its colour the pale brown of old straw. It lacks the bold leg stripes of *Theobaldia* but does have a few less obvious ones along its abdomen. *Culex pipiens* is probably our most abundant mosquito, occurring almost everywhere and breeding in even the tiniest accumulation of water. It is common in our towns and shares the habit of occasionally hibernating in human habitation; in cellars, for example, where it sometimes occurs in large numbers.

The use of the Underground for shelters had a footnote in 1948 when the deep-level tunnel at Clapham South was used as a temporary home for new arrivals on the *Windrush*. For several weeks over 200 Jamaicans were housed here and an underground labour exchange replaced the previous wartime facilities. This interlude aside, with the end of the war, the Tube had returned to its traditional use. The disappearance of the close-packed masses of sleeping humanity, with their rich odours and their radiant heat, would have hit the mosquitoes hard and reduced their numbers significantly. The fact that they did not disappear altogether was attested by the continuing accounts of maintenance staff and of passengers receiving unexpected bites. No one seems to have paid much attention to the phenomenon, however, until, in the late 1990s, it attracted the attention of geneticists working at the University of London.

It was around this time that Katharine Byrne, a researcher at the university, began to accompany maintenance workers on their nightly inspections of the tunnels, walking with them the stretches between stations and examining flooded sumps and shafts along the way. From seven different locations on the Central, Victoria and Bakerloo lines she collected water samples containing mosquito larvae and took them back to the laboratory to watch them develop. When the adult mosquitoes emerged, she took protein samples and, with her colleague Richard Nichols, subjected them to genetic analysis. The results were compared with a similar analysis of mosquitoes collected from a number of surface locations. When the results were published in 1999 in the journal *Heredity*, they were to give rise to considerable interest, particularly for the way in which they seemed to challenge conventional Darwinian theory.

The protein analysis conducted by Byrne and Nichols had revealed that there were distinct genetic differences between the Underground and the surface-dwelling mosquitoes. And yet they were, to all intents and purposes, physically indistinguishable. What separated them was not their morphology but their lifestyles. *Culex pipiens* is a mosquito that spends the winter in hibernation yet the population in the Underground was active all year. Living in the constant temperature and permanently frost-free environment of the tunnels they had simply dispensed with the winter sleep altogether. The constricted nature of the tunnels, and their permanent darkness, also meant that the usual practice of forming tall, dancing columns as a prelude to mating was no longer feasible. The mosquitoes had dispensed with this, too, and each male simply set out on its own individual pursuit of a female. Nor was this all. There is, worldwide, a small number of mosquito species that are 'autogenous', where the females, that is, do not actually need a blood feed immediately before egg laying but can store it up for later. The Underground *Culex pipiens* seemed to have joined their number. Not only did they not need the immediate blood feed but they had changed the prey from which they took it. The 'normal' surface populations of *Culex pipiens* fed from birds. It was, it seemed, a completely new adaptation of the Underground form to seek out a mammalian prey; the rats and mice of the Underground, and the humans.

Having set out these differences Byrne and Nichols hovered only briefly over the issue of how to classify this new variant. Others were to pursue the question with enthusiasm. Was the new Underground mosquito a different species, a subspecies, a semi-species or a physiological form? It is the sort of debate that taxonomists love, lining up on different sides to fire off a barrage of contradictory opinions as they attempt to impose some rigidity upon the shifting and often indefinable shape of reality. It is perhaps part of the natural human tendency to impose order, to reduce the endlessly creative universal chaos to a framework that is comprehensible. Byrne and Nichols simply side-stepped the matter, adopting for the London Underground mosquito the cumbersome designation of '*Culex pipiens* forma *molestus*'. Whatever the name, it is worth noting at this point that all the mosquito specimens collected by Byrne seem to have been of this single species. There is no mention of anything other than the *Culex pipiens molestus* hatching from the larvae that they had collected. It might come to be significant that *Theobaldia annulata* no longer seemed to exist as a denizen of the London Underground.

The final part of the Byrne and Nichols paper dealt with the possible origin of the new Underground form. How had it actually made its way there in the first place? There are, they reported, some populations of *Culex pipiens* in the Mediterranean that share some of the features of the London Underground population. Could they, therefore, have arisen from long-distance migrants arriving in this country and then finding their way into the tunnels? The paper dismisses the idea, concluding instead that the entire London Underground population arose from one single colonising event; that *Culex pipiens* mosquitoes entering, or trapped in, the Underground system had adapted, both genetically and behaviourally, to the conditions in which they found themselves. Moreover, they claimed, there is evidence that populations on the different Tube lines, being separated from each other by the action of trains, are themselves becoming distinct. There might, in other words, eventually be a Central Line mosquito, a Victoria Line mosquito and a Bakerloo Line mosquito. In the classical Darwinian model, evolution is taken to be a slow and cumulative process, happening by trial and error, bit by bit and over thousands of years. Yet here, the paper seemed to suggest, it was happening with startling rapidity, in the here and now, and literally before our eyes. It was this implication that made the paper so significant.

* * *

Despite its prevalence in the North Kent marshes, malaria was already in decline by the time that Dickens was writing *Great Expectations*. As an English phenomenon, it had reached its high tide in the mid-eighteenth century and thereafter began a slow withdrawal. Between 1840 and 1910, and despite the outbreak of several epidemics, deaths related to malaria in England were down to a total of 8,209. The decline of the disease has often been related to the widespread drainage of the English lowlands and the scale of this transformation of our countryside can hardly be overstated. According to Bankoff, as much as 50,000 square kilometres, or 12 million acres, was drained in the course of the nineteenth century, an area, he says, equivalent to half the agricultural land in England. The process hardly lost pace in the twentieth century, particularly in the big agricultural drive around the Second World War, when, for example, East Anglia lost 94% of its fenlands. Despite this, modern historians are cautious about ascribing the decline of malaria to this one cause. The same period saw other significant social changes all of which could have contributed to the decline of a malarial parasite that was already existing at the edge of its range. People lived in better ventilated homes and no longer in such close proximity to their livestock. At the same time, the overall number of cattle had increased dramatically, from 2.8 million in 1840 to over 6 million in 1910, providing an abundant alternative host for the mosquito. The widespread use of quinine as a malarial treatment had also weakened the disease and was increasing people's resistance. Most of all, living standards were rising, health and hygiene improving and advances being made in sanitation. It was this, more than anything else, that helped to turn the ague into history.

The malarial plasmodium had gone, but not so its carriers. Despite the scale of the drainage the Anopheline mosquitoes continued to breed in large numbers, in brackish coastal waters and saline pools. John Marshall, whose seminal work, *The British Mosquitoes*, I had handled at the British Library, had his own encounter with them when he moved, with his wife, to Hayling Island on England's south coast. It was an encounter that was to change the course of his life and to shape our future understanding of mosquitoes.

John Frederick Marshall CBE, MA, FLS, FRES, was born in London on 5 September 1874. Commonly known as Jack, he was the only child of Charles and Jennie, with a grandfather who had founded the Marshall and Snelgrove chain of department stores, later to merge with Debenhams. His fortune in life thus assured, he was sent to the famous Rugby School and then on to Cambridge where he achieved a First in Mechanical Sciences. He also seems to have studied law for he was called to the Bar in 1902, although he never actually practised. In that same year he married Emily Blanche Hughes of Chelsea. Blanche, as she preferred to be known, was of more exotic stock. She had an early involvement with the circus, then used her 'psychic powers' to gain employment as a fortune teller. She had clearly become the clairvoyant of choice for society for she was said to have a horse and carriage smarter than that of the famous Lily Langtry. An alliance between a practical engineer and a psychic socialite might seem unlikely, but it lasted for the rest of their lives.

Jack and Blanche set up home in Hayling Park Road, Croydon, naming the house Somerleyton, and it was here, in 1907, that their only child, Joan, was born. Travelling to Portsmouth one day, the couple noticed a signpost to Hayling Island. Perhaps it appealed to Blanche's psychic proclivities but they were intrigued by the similarity of the name to that of the street they were living in. They turned aside to investigate; and fell in love with Hayling Island. Its shape is something like that of a suspended Hammerhead shark, with a long straight coastline at its southern, head, end. Now the main resort area of the island, it was, in the early twentieth century, still undeveloped and the island as a whole, sparsely populated. When Jack inherited the family fortune in 1908 he purchased here, a six-acre plot including its own stretch of beach. There was already a cottage on the site but next to it they built their own much larger home, naming it 'Seacourt'.

One might assume with such a significant investment that Jack would have got to know a little more about the area in advance. But that, perhaps, is not the way of the wealthy and it seems to have come as some surprise to Jack and Blanche that the area was infested with mosquitoes. It was, they found, impossible to sit or work in their garden in the afternoon or evening, and at any outdoor events they had to swathe themselves in blankets. 'Either the mosquitoes go or I do,' Jack is reputed to have said, 'and I refuse to be driven out of my own house.' It was the mosquitoes that went.

Jack set about the task of dispatching them with typical energy. Hayling Island is flat and marshy and was then still undrained, dotted with salt flats, muddy channels and small, brackish pools, particularly in the 'Salternes', the oyster pools of its western and eastern fringes. Setting up the Hayling Mosquito Council, Jack enrolled the local population in mapping breeding locations and collecting thousands of specimens. He divided the island into nineteen 'mopping-up' areas, each with its own Secretary, and sent out dozens of volunteers to drain the pools or spray them with paraffin, or with White Cross disinfecting fluid which they bought from local shops. By 1924 the problem had almost entirely disappeared.

The success of the scheme built Jack's reputation and others, both at home and abroad, began to seek out his advice. As a result, and with no formal background in the field, he went on to found the British Malaria Control Institute, building it in the grounds of his own house and at his own expense. It was an honour that Sir Ronald Ross, the man who had earned a Nobel Prize for scientifically establishing the link between malaria and the mosquito, came down to formally open the Institute on 31 August 1925, and subsequently served on its Council. It went from strength to strength. With a role in education, advice, identification and the scientific study of mosquitoes, it received specimens from around the world and had, by the early 1930s, responded to enquiries from more than 11,000 localities and examined over 4,300 batches of mosquitoes. When it ran courses on mosquitoes and their treatment, the Colonial Office was among its clients, training up Oxbridge graduates before dispatching them to the African colonies. Jack, who was by all accounts an engaging speaker, gave talks wherever he was invited, including one to an audience of 1,100 schoolchildren in Portsmouth. The Institute's eleven rooms also included a Mosquito Museum which, within six years, had been visited by over 9,000 people. I would love to have been among them.

In all of this work Jack was supported by his colleague, John Staley. John had begun as the gardener at Seacourt but became a co-worker with Jack, and a loyal lieutenant, throughout all his mosquito studies. Being an engineer, Marshall was able to invent or develop some unique equipment for his laboratories, and for the photographic room. It included a rearing chamber, an automatic titrator, a six-foot camera and the Moscon Macrograph Microscope which was able to project microscopic images for demonstrations, drawing and photography. It was with this equipment that he provided all the illustrations for *The British Mosquitoes* on its publication in 1938. His work at the Institute was cited admiringly in the book's preface by the Keeper of Entomology at the British Museum. From having no previous experience in biology John Marshall had become the world's leading authority on mosquitoes.

Marshall is one of the great, under-recognised examples of eccentric English endeavour and I was keen to visit the place where he had lived and worked. Together with my sister Gill, who then lived in nearby Emsworth, I travelled to the south of Hayling Island where the Seacourt estate had been located. The buildings of the Institute have disappeared, replaced now by two rather anodyne blocks of flats in a style that might be characterised as Seaside Modern. The estate has been separated from the sea by the road known as Sea Front but the original cottage on the site is still there. Though not small it had clearly not been large enough for the Marshalls and dwarfing it next door, though half hidden behind high hedges of Griselinia and bay, was Seacourt itself. It is now a rather sprawling building to which wings and annexes appear to have been added at random. A mix of red brick, pebble dash and plaster, of gables and mock-Tudor beams, it is topped with a red tiled roof in which I could see a little dormer window in the Arts and Crafts style, so low that it appeared like a sleepy half-closed eye in imminent danger of dropping off to sleep. The house, I felt, had seen better days. Now divided into apartments, the paint was peeling on its beams and some of its tiles were slipping.

Despite its early successes the British Mosquito Institute experienced regular financial difficulties and by the outbreak of the war, and compounded by Jack's own monetary problems, these had become acute. With the navy commandeering the house during the war years, the Marshalls had evacuated to Bournemouth, while Staley struggled on alone in the Institute. When they moved back in 1946 it was over. Seacourt was sold and the Institute was closed. Having converted its buildings into a house, Jack and Blanche moved into it, renaming it Somerleyton after their original Croydon home. Jack was to suffer two strokes and wrestle with bouts of depression. With his physical and mental health failing he moved into a nursing home in Portsmouth where he died in December 1949. Blanche continued to live at Somerleyton until her death in 1964 at the age of 92. Employed by the local council, John Staley continued to work in mosquito control until his retirement aged 74.

* * *

The idea that the London Underground mosquito was demonstrating a remarkably rapid rate of 'speciation' was an exciting one and was, by the end of the twentieth century, catching the imagination of scientists and science writers alike. The development of this new species, wrote David Reznick in *The 'Origin' Then and Now*, took place entirely between the publication of the first edition of *The Origin of Species* and the collection of specimens in the mid-1990s, 'far less time to form a species, by orders of magnitude, than imagined by Darwin'. Peter Marren makes a similar point in *Bugs Britannica*: 'though the Underground was dug only a century ago, its resident mosquito is already on the way to becoming a new species, a process that according to Darwinian theory should take thousands of years.' Menno Schilthuizen is particularly enthusiastic in the introduction to his 2018 book, *Darwin Comes to Town*. He heard Byrne speak at an Edinburgh conference, he says, and,

> *even though her audience consisted of seasoned evolutionary biologists, she managed to thrill us all . . . We have been taught that evolution is a slow process, imperceptibly whittling species over millions of years – not something that could take place within the short timespan of human urban history. [The London Underground mosquito] drives home the fact that evolution is not only the stuff of dinosaurs and geological epochs. It can actually be observed here and now!*

Perhaps all this excitement was best summed up in an article in *Lateral*, the online journal of the Cultural Studies Association.

Written by Emily Gregg in September 2015 it was entitled 'The Fastest Show on Earth'.

I have no problem with challenges to orthodox Darwinian theory and would quite like to have come out on the same side of the case. But I didn't. I had encountered too many records and references that were leading me to an entirely different conclusion. In their 1998 paper, Byrne and Nichols acknowledge that there are populations of *Culex pipiens* in other parts of the world, especially in Egypt and the Middle East, that demonstrate the same propensities as the London Underground mosquito. They feed on mammals and the autogenous females have no need for a blood feed immediately before egg laying. There is an intriguing reference to the subject in a letter from John Marshall who doubted that such creatures could exist. His daughter, Joan, was, in 1927, in Italy, and whilst there was dutifully collecting mosquito specimens, packing them in cotton wool in matchboxes and sending them off to her father. To one of them she attached the note that it seemed to be a specimen of *Culex pipiens* and that it had been biting her. His response was somewhat severe:

> *C. pipiens under no circumstances will bite human beings. It is presumed that they take their blood meal from birds. You will doubtless remember, if you can spare the time to think of such matters, that although Staley spent several days stripped to the waist in a cage of C. pipiens, none of them could be persuaded to bite him. This is a scientific fact, so in future make your observations with more care.*

Apart from the implication that Joan is having far too frivolous a time in Italy, it is the faithful assistant and not Jack, one might note, who has spent several days semi-naked in a cage of mosquitoes. Jack Marshall's confidence in 'scientific fact' was in this instance misplaced. Within two years, continental workers had formally recorded a form of *Culex pipiens* that was autogenous and that took its blood from mammals. Whether he then apologised to Joan goes unrecorded.

Jack not only became aware of these autogenous mosquitoes, he also became convinced that they existed in Britain. In 1935, together with John Staley, he published a paper entitled 'Exhibition of "autogenous" characteristics by a British strain of *pipiens*' and by the time of *The British Mosquitoes* had incorporated them into the book. Though most textbooks, he writes, will tell you that *Culex pipiens* rarely attacks human beings, he has now acquired several records to the contrary. They were particularly associated with urban districts and had been found in Harwich, Hull, Epsom and Plumstead. There were also records from central London including hotels in Charing Cross and blocks of flats in Westminster. 'It is worthy of note', he says, 'that all the London cases occur within a limited area near Trafalgar Square.'

Someone else who had direct experience of these London mosquitoes was the malariologist and entomologist Percy George Shute. Addressing the Royal Entomological Society in 1951 he recalled an earlier encounter with them. It was 1923 and, like Marshall, he believed that *Culex pipiens* fed only on small birds 'and occasionally', he added, 'frogs'. He was surprised, therefore, to be asked to investigate a mosquito nuisance in the heart of London, and to find, 'walls and ceilings of certain bedrooms in the district containing dozens of *Culex* gorged with mammalian blood'. Extensive searches for breeding-grounds, he reports, were made over the next two years but without success.

All of these accounts, it should be noted, predate the Second World War, with mosquitoes with all the characteristics of the human-feeding Underground form not just already in existence but living in colonies in London. There is, however, no report I have encountered of anyone being bitten on the London

Underground prior to 1939. The idea expressed on some websites that mosquitoes became trapped in the tunnels when they were first being dug seems unlikely in the extreme. What is much more likely is that the masses pressed into the wartime Tube, the heat and the sweat and 'unbearable stench' that went with them, were just part of the rich cocktail of chemical signals being spread far and wide as an attractant to the neighbourhood mosquitoes. The crowds coming and going, and all the paraphernalia they carted with them, would also no doubt have assisted in the insect's physical transit. Perhaps they did not immediately breed there. A detail in one of the official reports of the time suggests they had more of a 'visitor' status, breeding outside the Underground systems in nearby puddles which, they advised, should be treated with 'a drop of paraffin or a weak solution of carbolic.' If the wind is in their favour, the report suggests, 'the newcomers find it easy to enter the tunnels'. Almost certainly these 'newcomers' would have included several different species; *Theobaldia annulata* as well as the ordinary, above-ground, form of *Culex pipiens* and among them, some of the *molestus* form, the human feeder already recorded as common in the area around Trafalgar Square.

With the end of the war and the departure of the shelterers, the other species would have died out, but not so *Culex pipiens* forma *molestus*. The London Underground mosquito was already ideally adapted to subterranean life. Far from these mosquitoes having evolved at breakneck speed in the tunnels, they had, I believe, arrived already equipped for a lifestyle in constant temperatures and restricted spaces and with a ready food supply of human beings supplemented by the rats and mice of the tunnels.

As far as we know they are down there still, but *molestus* colonies have also continued above ground. In both May and November 2017, the residents of Tudor Rose Court in Fann Street complained to their landlord, the City of London, that they were being attacked by mosquitoes. Their complaint began to work its way through the arcane machinery of City government and was raised at the Aldersgate Wardmote. The problem, the Alderman and the Common Councillors were told, may have arisen following the pressure washing of drains in the basement lightwells of flats. It was perhaps rather melodramatic that specimens were sent to Porton Down and its Medical Entomology and Zoonoses Emergency Response Department. The Department replied that they might be specimens of the London Underground mosquito but since the matter was hardly within their remit were able to offer no further advice. The Wardmote formally noted that 'the citizens of Tudor Rose Court . . . are sorely troubled by being bitten by mosquitoes' and referred the matter on to the Grand Court of Wardmote. It is a commentary on the survival of the above-ground *molestus* mosquito and on the continuing archaic ways of the City.

* * *

With the origin of the London Underground mosquito resolved, there remained a few final questions that had been hovering like hungry insects at the back of my mind. The first concerned the habits of the male mosquito. With all the attention on the blood-feasting female, no one seemed to have asked what the males were finding to eat. Where were their nectar or plant sap or rotten vegetative juices to be found in the world of the tunnels? It seemed, to me, as big a mystery as the female's switch from bird to mammalian blood but it had attracted little, if any, attention. I had read of occasional examples of gender fluidity in mosquitoes; males that become gynandromorphs, taking on the markings of the female. Some of these have also been seen taking human blood. Could such sexual change, I wondered, be at work in the London Underground mosquito?

A more fundamental question was whether, with climate change and rising temperatures, malaria might make a return to this country. A warming climate gives a longer period in which *Plasmodium vivax* can develop and, under the current scenario of climate change, it would, by 2030, be viable in southern parts of the country for four months of every year. By 2080 it would be viable in Scotland. Given the other factors associated with the decline of malaria, however, the likelihood of its return remains slim and the real concern, in a warming world, is over the impact of new, invasive species and the diseases that might accompany them. *Culex modestus* (not to be confused with *molestus*) has recently been found in the country after an absence of eighty years and is known as a vector of West Nile disease. Also arriving is *Aedes albopictus*, a mosquito known from its size and pattern as the Asian tiger. It is a carrier of dengue, chikungunya and zika virus and has spread outwards from Asia, already reaching north America and as many as twenty-eight countries in Europe. In England it has been found breeding at different sites in Kent in every year since 2016, and in 2019 it appeared in London. Like so many other forms of wildlife, its spread has been associated with transport systems, being carried from place to place in the cargoes of lorries, especially those of car tyres. Its progress along international road networks could even be mapped by its appearance in motorway service stations. The United Kingdom has a 'National Contingency Plan for Invasive Mosquitoes' with its latest iteration May 2020. 'Modelling studies of likely spread or incursion of vector-borne diseases', it states, 'show that the problem of emerging or re-emerging vector-borne diseases may intensify and spread to the UK soon.' The plan defines five levels of risk, the lowest 0, the highest 4. The current risk level is assessed as 1.

Which all leads to my final, and shamefully anthropocentric, question: what is the point of mosquitoes? The question has been addressed in myths and legends across the ages, often with an answer that challenges the nature of power, with this tiny insect, this 'translucent phantom shred of a frail corpus', as Lawrence called it, able to bring down the mightiest of beasts. Timothy Winegard answers in a different way arguing that, 'while the mosquito is miraculously adaptable, it is also a purely narcissistic creature ... She has no purpose other than to propagate her species and perhaps to kill humans'. Could this, he wonders, be the ultimate Malthusian check of uncontrolled human population growth? With such a chilling idea, the mosquito's allegedly murderous intent becomes a sort of ecological balancing act. Others, however, have taken the contrary view, arguing that mosquitoes should be eliminated altogether. The idea was aired again in January 2016 in the *BBC News Magazine* asking, 'Would it be wrong to eradicate mosquitoes?' Among the scientists interviewed by journalist and radio presenter Claire Bates was Olivia Judson who supported what she described as 'the ultimate swatting', the elimination of thirty species of mosquito. It would, she claimed, save one million lives and only decrease the genetic diversity of the mosquito family by 1%. Others pointed out the role that the mosquito plays as a pollinator and as an important part of the food chain for birds, bats, fish and amphibians. Judson dismissed such ideas with the comment, 'We're not left with a wasteland every time a species vanishes,' arguing that other insects would soon step in to fill the ecological niche. It seems to me an argument of dangerous hubris; an arrant presumption that we can actually predict, in the inadequately understood and complex inter-relationships of ecology, the impacts of eliminating entire species. The world is full of unintended consequences. Moreover, the overall abundance of insects is already dropping so dramatically it is folly to think that other species are waiting in the wings to fill the mosquito's ecological niche. It is other approaches that will need to be used in combatting malaria and the good news, as I write this, is that the first malaria vaccinations for children are just being introduced.

It might not much mitigate their malarial role but I would like to say two small things in the mosquito's defence. The first is that, however bad its malarial impact, it has been much magnified by actions of our own; by our lifestyles, our environmental interventions, the movements of our armies, our hunger for colonies and now, increasingly, by international travel and by global trade. Even our haphazard use of medicines and our indiscriminate use of insecticides, some scientists have argued, has played a negative part. 'We reap what we sow,' as Winegard put it. 'Or where we sow, the Reaper appears.' The second argument for the defence is that the mosquito, far from having a 'purpose' of spreading malaria, is itself driven by its plasmodium parasite. Research has shown that the plasmodium is able to supress the female's ability to produce anticoagulant when feeding. It is a manipulation that reduces her intake at each feed, thus forcing her to bite more people, more frequently, thus more effectively spreading the virus. Once in the human bloodstream, the plasmodium can also transmit a 'bite me' signal attracting other mosquitoes and helping to complete its reproductive cycle. Even the radiant heat of a malarial fever might serve to attract more mosquitoes and aid the parasite's spread. The malarial plasmodium is, in its own right, a remarkable example of adaptive evolution, shaping the behaviour of the mosquito to suit its own needs.

Despite more than a century of intensive research, of which John Marshall was but a part, there is still much about the mosquito that remains unresolved. As Andrew Spielman points out in his book *Mosquito*, we still don't know why a blood feed is necessary for reproduction, how the insect distinguishes among its hosts or even how it has the necessary force to suck blood 'through a feeding tube with a diameter that is so infinitesimal that the resulting friction should make it impossible . . . The beauty of the mosquito lies in these mysteries and many others.'

I had come to the story of the mosquito through my attraction to the ambivalent lure of the subterranean, and I had found in it a creature that was unwittingly responsible for the death of billions but that could still be described in terms of its 'beauty'. It is an insect of awesome adaptation and of perfect form, and which, in its multiple forms, is thriving both above and below ground. It has helped shape history and defined a whole area of eastern England, and despite the massive attention we have paid it, many of its secrets remain unrevealed. It is in the face of such mystery that I find myself recalling a passage from the apocryphal Book of Wisdom. Take it as you will:

For you love all things that exist
And detest none of the things that you have made
For you would not have made anything if you had hated it.
How would anything have endured if you had not willed it?
Or how would anything not called forth by you have been
    preserved?
You spare all things, for they are yours, O Lord, you who love
    the living.

5

The Intergalactic Bear

'They are the most fascinating animals known to science', states the website of the American zoologist Paul Bartels, 'these poorly studied creatures have completely won me over.' The creatures in question are the microscopic animals known as tardigrades and his almost unscientific ardour seems to have been shared by all who have encountered them, from the German naturalist Gustav Jäger in 1867, who thought them 'the most miraculous of creatures', to the contemporary American entomologist Frank Roman III, who regarded them as living in a 'twilight zone', 'somewhere between reality and myth'.

For introducing me to them, I have to thank my youngest son, Tom. It was his first year at secondary school and his biology teacher was one of those whose enthusiasm for their subject both infects and inspires their charges. Coming home excitedly one day, he shared what he had learnt about the creatures colloquially known as 'water bears'. I read all the material the school had given him and went on to search out more. Like so many before me I was both enamoured and intrigued by these innocuous-seeming creatures whose existence poses so many difficult questions.

One of the great uncertainties surrounding the tardigrade is where, in evolutionary terms, it actually comes from. It seems impossible to classify, displaying features of many other animals but bearing a close relationship to none, as though it has been pasted together from a catalogue of parts. It stands so alone in the complexity of creation that, despite the relatively small number of its species, it has been afforded a phylum, the highest division in the classification of the animal kingdom. This peculiarity is compounded by another; its remarkable ability to withstand extremes, whether of heat, cold, desiccation, salinity, atmospheric pressure, toxic immersion or solar radiation, and not just for days and months but even years. Add to this its vanishingly scarce appearance in the fossil record and you have a combination of factors that is both puzzling and potent. The tardigrade seems to exist in a realm somewhere between science and science fiction. Or perhaps between fact and faith, its study being sometimes as theological as it was scientific. It has led to a debate on life after apparent death, to a claim that it disproved Darwin and to the suggestion by some scientists that its origins are extra-terrestrial; that the 'aliens', whose arrival we have so long, and so anxiously, awaited, are already here. What then is the real identity of the tardigrade and how does it come to be here? It was to answer these questions that I set out to investigate a creature I had never yet seen, but that was invisibly abundant about us and for which such extravagant claims were made.

* * *

Antonie van Leeuwenhoek was a draper, living, in the mid-seventeenth century in the Dutch town of Delft. Concerned to better examine the quality of the thread in the cloths in which he dealt, he developed an interest in lens-making, inventing new techniques which he was at some pains to keep secret. He was soon to turn his new and more powerful single-lens microscopes, not just to his trade, but to the natural world, looking at the anatomy of lice, the pattern of veins on a fly's wings, the crystals that form in the affliction of gout, the structure of red blood cells and spermatozoa. His revelations about the creatures that abound in a drop of pond water, creatures which he christened 'animalcules', was ridiculed by many, unable to believe that a drop of clear water, the same substance that they daily drank, could be teeming with life.

Leeuwenhoek was opening up a world of previously unimagined wonder. Prior to his work, the use of magnifying lenses had been turned almost entirely in the other direction; towards the skies. The telescope had enabled the study of planets and their moons and of the closest systems of stars. It had begun a journey that would reveal the unimaginable vastness of a universe where galaxies are numbered in their billions, and planets in their trillions, and where the impossible becomes probable in dark energy, black holes, curved time, multiple dimensions, parallel universes and nature's 'fifth force'. The new advances in the technology of the microscope were now beginning a parallel journey into intimate rather than overarching space; one that would lead eventually to the discovery of smaller and smaller organisms and to the vastness of the minute, of atomic and subatomic particles, of protons, neutrons, electrons, neutrinos, leptons, muons, the Higgs boson and five types of quark. We are becoming less and less significant as space moves away from us in both directions, with things becoming both too big and too small to understand.

The British scientist Robert Hooke, a contemporary of Leeuwenhoek, began illustrating the wonderful new forms the microscopists were encountering and published them in his book *Micrographia*, perhaps the first scientific bestseller. Here were water fleas like the cyclops, shaped like a tear-drop but with forking tail and a single eye, or the rotund daphnia, floating upright like a manatee but fitted with feathery limbs that project ornately above its head like an Ascot-goer's hat. Here, too, were diatoms like intricate baubles of glass, the hydra, a shape-shifting tube with a tentacled head, the slipper-shaped paramecium, fringed with waving hair-like filaments, and the formless amoeba, pushing out its plasmodium as it flows and slithers and shifts. No wonder so many derived their names from Greek myth; they were every bit as fantastical as the creatures that had adorned the pages of the early bestiaries or decorated the corners of ancient maps.

Into this world appeared the tardigrade. The clergyman Johann Eichhorn insisted he had observed them first, in 1767, but he failed to mention the fact for another fifteen years when retrospectively, and rather grumpily, he attempted to claim precedence. Even then he described the creature rather dismissively, saying it had 'nothing which might make it attractive to the eye' and compounded the insult by illustrating it with the wrong number of legs. It seems fair, therefore, to give priority to Johann August Ephraim Goeze. Goeze was another clergyman and pastor of the church of St Blasii in the German town of Quedlinburg. First observing the tardigrade in 1773 he made extensive notes, describing it as the 'most rare' of creatures, which it isn't, and 'the most strange', which it is. His published description covers its appearance and many aspects of its behaviour, including its feeding habits, and he concluded that it was some kind of worm. As might be thought appropriate for a Pastor, his description ends with a pious but lyrical passage of praise:

Creator of the elephants and atoms, of the whales and small living points in water! I am astonished by the endless variety of designs, according to which your wisdom has formed in a different way the body of each animal; the bird, the frog, the insect and the worm! . . . The almighty God said: Be it! [and] this drop in a bucket ran out of his hand . . . Lord! Who has been your advisor? He has created everything, the sun, the clouds, the oceans, the depths, the visible and invisible worlds, the big wild animal and the worm unseen.

It was Goeze who coined the common name for this 'worm unseen', calling it the '*kleine wasserbär*': 'Strange is this little animal, because of its exceptional and strange morphology and because it closely resembles a bear in miniature. That is why I decided to call it "little water bear".' The name caught on, though it was perhaps as much to do with its lumbering gait as with its appearance. Among the many who continued with the bear analogy was the 'Honourable Mrs Ward' in her book, *The Microscope*, published in 1867 and a classic of its time. She describes a first encounter with the species in which it seems to stare at her for minutes on end, 'in that position not a little resembling the white polar bear'. Much later, the ursine connection would be taken further still with the naming of a particular new species. To this day, only three fossil specimens of the tardigrade have been found and when the first was named in 1967 it was an event of some significance. It was discovered in a collection of amber, taken by William Legg some twenty-one years earlier from beside the Saskatchewan River in Manitoba. Legg had long since died before his mentor, Kenneth W. Cooper, examined the amber and identified in it the creature he named *Beorn leggi*. The *leggi* was in his friend's honour, the *Beorn* came from a character in *The Hobbit*; 'a huge man with a thick black beard and hair, and great arms and legs with knotted muscles'. He was a man who also happened to turn into a bear at night.

* * *

Personally I am unconvinced by the bear analogy. To me the tardigrade is rather more cigar-shaped, resembling, perhaps, a miniaturised Zeppelin airship, but one with the addition of legs. Though a tiny creature, it is at the larger end of the microscopic scale, something under a millimetre in length. An oft-repeated image on the internet shows it with projecting snout and crinkled, baggy skin, like an ill-fitting clown suit. This thick but flexible cuticle, as folded as that of the Indian rhinoceros, will be shed several times during the animal's life, with a new one growing beneath it each time. Some species will lay their eggs inside the shed skin, giving them a degree of protection from predators and from the dangers of drying out. A few others have no colon and along with their skin will deposit all their accumulated defecation in one go. The image shows it the colour of military khaki but internet pictures should be treated with caution; computer animation can misrepresent a creature's movements while images from electron microscopes usually add colour. In more sober reality, many tardigrades are completely transparent, any colour in them coming from the visible contents of their gut. Their bodies are segmented, though the segments are hard to detect, with a head, three main segments and final or caudal section. From the sides of the body sprout three of its four pairs of curious, unjointed and stubby legs, each ending in a tuft of hooked claws that have been described as resembling a 'fistful of mediaeval weapons'. They do not swim, as many of the images seem to suggest, but make a rather bumbling progress that has earned them the scientific name of *Tardigrada*. Concocted from Latin by the Italian naturalist Lazzaro Spallanzani in 1777, it literally means 'slow-paced'. Apparently he was thinking, at the time, of a tortoise. What does not contribute much to its progress is the fourth pair of legs. Unlike those of any other animal, they are attached to the animal's rear and project from it backwards. With them it will anchor itself onto something, when eating perhaps, or when attempting a difficult manoeuvre. F.J.W.

Plaskitt's 1926 volume on *Microscopic Fresh Water Life*, another biological classic, includes his own description of the animal's overall movement. Although a serious scientist in every other respect, the tardigrade leads him, as it had so many others, into imaginative prose and a surprising, if rather refreshing, degree of anthropomorphism:

> *Altogether they are very interesting creatures to watch, and even comical at times as they persist in trying to catch hold of objects that are not there. They appear quite unable to profit by experience and will continue clawing downwards and forwards upon a slippery surface or in the open water, seemingly to obtain a foothold, until, rather agitated, there is a final rally and quickening pace, without result, followed by a literal shake of the head and an expressive turn of the body, evincing their evident disgust of things and probably of a Water Bear's life in general.*

I do not think that tardigrades need to be too disgusted about their lives as they move rather placidly about the thin film of water in which they live. Their explorations are aided, in many species, by rudimentary eyes, one on either side of the head, though in practice they are simple structures and not much more than pigmented light-detecting spots. Among other similarities to 'higher' species they also have a rudimentary nervous system and a long gut, though they lack lungs or any other respiratory organs, having the ability to absorb oxygen from all over the body surface. A curious feature of the anatomy, sometimes known as 'mystery cells', are the coelomycetes, small, spherical objects that are apparently unattached to anything and swill about in the body cavity as the animal moves. Their actual number varies greatly between individuals, even within a single individual's lifespan, and their function remains unclear. Almost as curious is the tardigrade mouth. It is tube-like and leads to a pharynx that works like the bulb on an old-fashioned perfume bottle, expanding and contracting to draw in a flow of fluid. Concealed and sheathed within the mouth are two sharp, pointed instruments known as stylets. When the animal is feeding they are exerted through the mouth to pierce the tissue of plants or prey. When the tardigrade sheds its skin it will shed these stylets along with it before growing a new pair, which seems rather wasteful, unless, perhaps, it is a way of keeping them effective and sharp. It does, after all, depend on them for its survival.

The food of the tardigrade consists of plant cells, fungal spores or algae. The preferred food of particular species is reflected in subtle differences in the spacing of the stylets, with some species being carnivorous and eating such other small organisms as rotifers, nematodes and protozoa, and occasionally other tardigrades. Pastor Goeze was clearly quite excited by this attribute describing them as 'beasts of prey of the invisible world'. Nonetheless, we shouldn't be afraid to watch them, he reassures us, in a tone which seems concerned with protecting our sensibility: 'they are beasts of prey in relation to the other animalcules of their worlds, similar to the tigers and the lions of the African deserts . . . In this way our world is organised'.

* * *

The 'way our world is organised', it turns out, may not be quite so straightforward, certainly as far as the tardigrade is concerned. '[Its] evolutionary history', wrote Martin Mach in the online *Baertierchen* journal, 'is largely a mystery'. Having contributed as many as 238 articles on the creature he is probably well placed to know. Biologists have wrestled with taxonomic uncertainty ever since the animal was discovered. It sits uneasily within the accepted scheme of things and compounds this by questioning whether the schema is correct in the first place. Here is an animal that is segmented, like the Annelida, that great phylum of living things to which the earthworms belong. It has piercing mouthparts like the vast group of roundworms, the unsegmented nematodes. It has eight legs like the spiders, scorpions and crabs, but legs that, unlike theirs, are completely unjointed. It is fleshy and lobe-like, and in this respect resembles the *Onychophora*, another peculiar and problematic phylum of stubby-legged creatures otherwise known as velvet worms. And it sheds its skin in a way that parallels the insects and other Arthropods, the spiders, scorpions and crabs. Trying to find a 'right' place for it is like trying to put a biological label on the Chimaera.

When Goeze wrote up his first observations of tardigrades they were published under the title 'Some Observations on Peculiar Water Insects'. His description of them as 'worms', and the possibility that they may be related to one or other of the various worm-like groups, reappears regularly over the years that follow. The majority of early opinion, however, was that they were related to the rotifers, another group of the water-living animalcules that had been studied by Leeuwenhoek and that were themselves of uncertain taxonomy. When Linnaeus introduced his now universal binomial system of classification, the first person to apply it to the tardigrades was the Danish naturalist Frederick Muller. He considered them to be related to the *Acari*, or mites, those small relatives of the spiders that often turn up in infestations of the house, the garden or the chicken run. For more than two hundred years the argument went to and fro with different affiliations being argued on the basis of different parts of the animal's anatomy: its pharynx, its stylets, its legs, its claws, its skin, its sense organs, its muscles, its internal organs and even its ability to withstand desiccation. Depending on where the particular researcher's focus lay, they were ascribed to rotifers, worms, insects, spiders, mites or the crustaceans. Some early nineteenth-century naturalists did not regard them as animals in their own right at all, but rather as the larvae of one or other group, of fleas or beetles or mites. Mrs Ward was one of those who came down on the Arachnid side of the argument, rejecting the still current rotifer theory and suggesting that they were 'physiologically speaking, poor relations of the great family of spiders'. F.J.W. Plaskitt was of a similar view, relegating them, along with other problematic organisms, to the final 'Miscellaneous' chapter of his book, then suggesting an affinity to the spiders and mites within which order they would, he said, represent the 'lowest section'. Notwithstanding all the learned argument and detailed research, their relationship remains unresolved and two more recent papers have added further pieces to the puzzle. In 1997, American scientists working on DNA sequencing, suggested that all species of invertebrate which periodically shed their skin are genetically related. Thus the tiny worm-like nematodes, for example, would be more closely related to insects and crabs than they would be to earthworms or flatworms. They christened this new 'superphylum' Ecdysozoa, a grouping that would include more than four and a half million species, including the tardigrades. A further twist in the tale came with work in 2018 suggesting that tardigrades themselves had become 'miniaturised' during their early evolutionary history. Extensive modification from the earliest forms had led to a loss of some of their genes and a reduction in the number of cells in all their organs. Their bodies had grown smaller, losing several segments as well as their circulatory and respiratory systems. No fossil evidence has yet been found of such creatures, but combining these two sets of conclusions must surely raise the possibility that these much larger one-time tardigrades were the common ancestors of many of our modern invertebrates.

Throughout all the debate, the animals have remained firmly fixed in a phylum of their own, the *Tardigrada*, the name first given to them by Spallanzani. The significance of being elevated to phylum status cannot be underestimated. The total number of phyla in the entire animal kingdom is generally put at thirty-five, containing between them many millions of species. That one group of just over 1,000 species should have a phylum all to itself is a matter of rare distinction. Or, perhaps, of splendid isolation.

* * *

To get to know these inhabitants of what Goeze called the 'invisible world', was clearly going to present me with some problems, not least my lack of experience with the microscope. Here were difficulties of a different order from travelling to see an ancient yew or seeking out a lichen or musk. I had, in my researches so far, already encountered *The Biology of Tardigrades* by Ian Kinchin, and had read it from cover to cover. Written in 1994, it seemed to me still the most comprehensive treatment of the creatures and having tracked down its author I was pleased when he readily agreed to meet. I encountered the virtual Ian via Zoom, sitting in what I took to be his study, in front of several rows of white, book-laden shelves. Ian is based at the University of Surrey and I had been somewhat surprised to find that he is a Professor of Higher Education rather than something senior in biology or invertebrate zoology. I asked him, therefore, about his personal journey into tardigradology. His first degree, he told me, had been in biology, and when he was looking for a subject for his final dissertation his supervisor had suggested tardigrades, on the basis that nobody knew anything much about them. With the assurance that whatever he came up with would probably be new, Ian had agreed, and so the relationship began. Ian had gone on to become a science teacher but his interest in tardigrades continued and some sixteen years later he took a year's sabbatical to complete a master's degree on the subject. 'After all,' he told me, 'playing with electron microscopes was always a lot of fun'. He had also used this time to write his book. 'Tardigrades were regarded as a fairly minor group with no medical or agricultural implications, so there was simply no whole book in English on the subject at that time.' Ian had stepped forward to fill the gap. Apart, perhaps, from Hooke's *Micrographia*, books on invertebrate zoology tend not to be great earners and it soon became clear that 'tardigrades were not going to pay the mortgage'. He had returned to a career in education and was now a Professor training other university lecturers to teach, rather than 'just read out their PowerPoint slides'. Biology, meanwhile, he lamented, had disappeared into the more and more technical world of gene sequencing and although he still maintained an arm's-length interest in tardigrades, he wondered, rather diffidently, how much he would be able to help me. I felt rather confident that he would.

I asked him about the 'awe and affection' with which, since their very discovery, serious researchers seemed to regard these animals. It was something he readily acknowledged. 'When you first see them you just make these "ah" and "oh" noises. They are so amusing to watch,' he said, amusement, I noted, being the tenor of several of the early accounts. 'They are so unlike anything else of that scale, they're really different . . . there's so much going on with them that you end up sitting and watching them and an hour's gone by, and you think, wow, I've done nothing but sit and watch this thing because it's so fascinating . . . people tend to get carried away and a bit poetic when they see them.'

He was right about the poetry. Researching *Beorn leggi* I had already discovered that the later discovery of a third type of fossil tardigrade had led one of the researchers to compose a song, one which I had heard and enjoyed on YouTube. I had also come across a song composed and recorded by Cosmo Sheldrake in which a tardigrade, aware of its powers of survival, wrestles with the question of whether to leave its simple home and head for greater things. Several verses of self-doubt later it decides to stay just where it is, in its home in the shrubbery with a cabinet full of whisky and its feet, rather strangely, clad in socks. From

Ian, on the other hand, I was accumulating more accurate information about the creature, as he talked enthusiastically about its structure, its resilience and its ecology, which, to my disappointment, did not seem to include whisky. It was only towards the end of our interview he revealed that he had discovered a new species himself. He had found it, moreover, in the gutter of his own flat in Guildford, and in considerable numbers. He had sent some to a taxonomic expert in Italy who confirmed that it was indeed something previously undiscovered and he had made it the subject of his master's degree. 'When I did my viva my examiner raised an eyebrow and said, "You're telling me you're doing your master's degree on a new animal that only exists in the gutter of the flat where you live?" "No," I answered, "it's in the garden too."'

Despite this scepticism, they had passed him.

Ian then had the job of naming the new species and together with Roberto Bertolani, his Italian collaborator, they came up with the rather uneuphonious epithet of *Ramazzottius varieornatus*. Since that was the genus to which it belonged, the *Ramazzottius* part of the name was a given, but I asked about the *varieornatus*. The structure of a tardigrade's eggs, I learnt, is an important element in determining its identity. Some tardigrades are parthenogenetic, meaning that, like my pet stick insects, the females can breed without the help of a male. Some species are hermaphroditic, but the majority breed bisexually. However they produce them, their eggs are beautifully sculpted. 'They are just fantastic', enthused Ian, 'amazing structures, so delicate and so intricate for something that is so small.' Some display extraordinary symmetries while others have tremendous geometric complexity. Some are shaped like opening flowers, while some are decorated with spines and projections. It is through this diversity of egg shape, rather than by features of the adult animal, that many tardigrade species are determined. Bertolani had discovered that the eggs of this new form were

ornamented with spikes, hairs and whiskers that came in varying sizes; hence *varieornatus*.

I remembered that in his book Ian had speculated on the
reason for these strange sculptings. They might help anchor the
eggs on a surface or provide a defence against predators. They
might help in the regulation of gas exchange or perhaps delay
the process of desiccation. In our conversation, it was the last of
these reasons that he stressed, suggesting that the spines or
structures trap water, providing a reservoir that slows down
drying in times of drought.

Perhaps I am easy to impress but I thought it a rather grand
thing to have found and named a new species. But Ian, again,
was rather diffident. 'Of course it's always exciting to find
something new; great, fine, very happy . . . but it doesn't change
the world, it's not a cure for cancer, it's just a very little thing,
most people would probably think, "Who cares?" At the time I
was quite chuffed . . . but whether it contributes to science or
not is for others to decide.' I ventured to suggest that we didn't
always know which discoveries might have a future significance.
In fact I was later to read that Ian's own species had been
pressed to the service of humanity. A team at the University of
Tokyo had studied *Ramazzottius varieornatus* and found that it
contained a protein, unique to tardigrades, that protected its
DNA against damage from X-rays. Applying this same protein
to cultured human cells, they found that it reduced X-ray
damage by around 40%. The substance could, they hoped, have
a use in protecting those who regularly worked with X-rays.
Ian, however, had been clear that naming things was not the
important issue:

> The classification and naming of numerous species is a bit like
> stamp collecting. The difference between one species and
> another can be so small . . . this claw goes one way, this claw
> goes another, so it's a new species . . . Well OK, but are they

different species or just variants on a theme? And at the end of the day, does it matter whether there are 720 species or 750? That's not the interesting thing . . . it's their biology that's interesting, their biochemistry that's so fascinating, their ecology, how they fit into their ecosystem . . . And all this stuff is in your back garden. There is a whole discussion now about what they call the 'extinction of experience' because people are removed from nature and it's a great loss, but this stuff is in your back yard, and if you live in a flat, it's on your balcony, millions of them probably, that you're just not aware of. And these things are so complex, with eyes and mouths and nervous systems all in the space of a tenth of a millimetre. How could you not get excited by that!

How indeed.

* * *

When Plaskitt wrote his affectionate portrait of Tardigrades in 1926 he was aware of just one British species and not many more than twenty had been identified worldwide. He correctly predicted that more would be found, and Ian's back-yard discovery of *Ramazzottius varieornatus* was just one part of an explosion in species that was to occur over the next hundred years. Today the worldwide tally has reached a figure somewhere between 1,300 and 1,500, a numerical uncertainty that rather justifies Ian's scepticism on the subject of naming. Using diversity estimation techniques Paul Bartels has suggested the total number of species could eventually reach 2,700. It seems plenty enough, but for a whole phylum it is, in fact, surprisingly low. Among these species two basic forms of tardigrade have been determined. The Eutardigrades are smooth-skinned, rather caterpillar-like and lacking in sensory receptors; the Heterotardigrades more resemble miniature armadillos, with complex armature often bearing protrusions, spines or long hair-like filaments. Though largely land-based, there are some ocean-dwelling species among them, and they come in forms that could be regarded as the product of a drug-induced dream; covered in tubercles, dressed up like a Japanese samurai or trailing long, decorative appendages like an aqueous bird of paradise.

This profusion of tardigrades is, we now know, global in its distribution. They occur in every continent and in every country, the German zoologist Eveline du Bois-Reymond Marcus even reporting one from an Oceanic island some 2,000 kilometres from the nearest large land mass. They had reached there, she believed, by floating on mats of sargassum seaweed. From the poles to the equator, wherever there is a modicum of moisture, there will also be tardigrades. They are found on mountain tops and in deep-sea sediments, in hot springs and mud volcanoes. They occur in the tops of the tallest trees as well as in the leaf litter that gathers beneath them. In freshwater sediments they can number as many as 25,000 per litre. One species is found in the body cavities of barnacles, another within the shells of mussels. They even survive beneath solid ice. In 1998, the French explorer and 'glacionaut', Janot Lamberton, made the deepest ever descent into an ice cave, formed by melt water during the Arctic summer, a rift or moulin of some 202 metres. From it he brought back tardigrades.

The number of British species has been put at seventy-four. The records seem to show a strange absence from the east of the country, a fact which maps the distribution of tardigrade collectors more than that of tardigrades. With this paucity of interested parties over much of the country, and with such great gaps in our knowledge, it seemed to me that there might still be a role for the enthusiastic amateur. I was not sure I had the commitment to become a total tardigradologist, but I did, at least, want to dip my toe in the pond water. First of all, however, I had to find them. In all the literature there is one habitat that is referred to repeatedly. The tardigrade, it seems, is most at home in a moss. 'If I had to choose one animal whose life is most closely tied to the life of mosses', wrote Robin Wall Kimmerer in her book *Gathering Moss*, 'it would be the Water bear.' In her lyrical account of this group of plants she relates a study counting the number of organisms found in one gram of moss from the forest floor, a piece 'about the size of a muffin'. It included 150,000 assorted protozoa, 3,000 springtails, 800 rotifers and smaller numbers of nematodes, mites and fly larvae. There were also 132,000 tardigrades. It is no wonder that another common name of the tardigrade is the 'moss piglet'.

For me, this posed a problem. Along with sedges, fungus gnats and the microspecies of bramble, mosses are one of the areas of natural history into which I have rarely ventured. Birds and wild flowers I have looked at for much of my life. Many of them I recognise with no conscious effort, identified by a sort of general unspecific impression, what bird-watchers call a 'jizz', that prompts a name to rise unbidden from my overcrowded cortex. Most common trees I can recognise from a speeding train, their identity revealed by shape or silhouette alone. But mosses are a different matter and with them I was back in the naturalist's nursery class. Knowing that they were a favoured home of the tardigrade I began to take more notice of them, an exercise I thought of as 'sharpening my moss awareness'. On daily walks I would scan the tops of walls, the base of buildings, the cracks beside a kerb stone, the damp bricks behind a leaking gutter or the bare patches of lawns on estates. In unsavoury places I would attract suspicious looks from passers-by as I bent down to scrape off samples and stuff them into one of the used envelopes I was carrying in my pocket. Struggling to determine each species – and even to determine whether each sample was in fact something different – I developed a much greater sympathy for those on my walks who had wrestled with the difference between cow parsley and hemlock or dunnock and sparrow. To name something can be to limit it and learning a label can short-circuit a journey of knowledge. But it can also be a beginning. To call something by name is also the start of a relationship; it is to show a proper respect.

Gradually, then, I was getting to know these neighbours. There was one that appeared almost everywhere; on soil, paths, roads, walls, waste ground and roofs. It seemed to particularly enjoy the edges of things. Here it formed dense thatches, flat and compact and almost emerald green. The shoots were so tiny that, individually, they were hardly discernible and packed together gave the impression of a single velvet sheet, the baize perhaps of a billiard table. Under my lens they were dotted with sparkling, glistening pits. This was *Bryum argenteum*, the silver-moss, its silvery sheen becoming more apparent as it dried. As a lover of nitrogen it is at home in our polluted and dog-infested cities, but before our conurbations it would have been rarer, concentrated beneath bat roosts or on the guano of colonies of sea-birds. The city, in other words, is its artificial midden.

There were others with which I was soon on first-name terms, the affirmation of their existence seeming to multiply my own. *Tortula muralis*, the wall screw moss, formed bright green and neatly rounded cushions that balanced themselves on the tops of walls. From these cushions rose multiple thin stems, topped with tapering spore capsules, green but burnished red at the tips. Viewed from eye level they resembled the massed minarets of an ancient oriental city. Common, too, was *Grimmia pulvinata*, the grey-cushioned grimmia, which formed furry cushions, their frosted appearance arising from the long greyish hairs that project far beyond the tips of their tapering leaves and form, under the lens, a great gleaming tangle. In the car park of my local Tesco was *Ceratodon purpureus*, or redshank, distinguished by its close-packed mass of red fruiting stems. I knew this place well, both as a naturalist and a shopper. I had once seen a flock of siskin here, feeding on the Italian alder, and it was now home to our single local collared dove with its melancholic, unmated calling. But as for this claret-wine stain on the concrete, I must have walked past a dozen times, without seeing it.

The 'Urban Cliff' hypothesis suggests that there is a parallel between the plant and animal species found on cliffs and rocks and those which grow on the vertical surfaces of our cities. It also suggests that this relationship has been going on since humans first made their homes in caves and rock shelters. You do not need the support of a hypothesis, however, to be familiar with the idea; it is apparent in many of our urban birds, for example, the feral pigeon and the peregrine that feeds upon it, the kestrel, the house martin and the swift, all of them originally nesting on sea cliffs or rocky slopes or stony ledges. It can be seen, too, in city plants; the ubiquitous buddleia, the ivy-leaved toadflax and the pellitory-of-the-wall. It is there, too, in our urban mosses which have made these artificial brick and stone and tarmac surfaces their home.

Having observed this on our 'urban cliffs', it would be good, I thought, to visit the rural original; a natural rock face. I turned again to Dave who, in a sort of revolving door of natural history passions, adopts a different specialism each year, whether it is bark-burrowing beetles, woodland ferns or foliate lichen. I have known him spend whole summers mapping the nests of wood ants or tracking the spawning grounds of brook trout. By serendipitous chance his current focus was on mosses and, in particular, he was surveying the mosses of the Wealden sandstone ridge. It was an opportunity I could not miss, and he agreed to take me with him, though I was, I suspect, more of an encumbrance than any sort of help.

Our target was an outcrop of the Tonbridge Wells sandstone that formed the highest ridge in the Weald. Below us, the river valley was invisible in the morning mist, a thick white gloom that had formed in the night and then grown so weary that it had sunk down into the hollows, leaving the rest of the day to break bright and clear above it. The ridge, a Site of Special

Scientific Interest designated largely for its mosses, is set in the woodland of Eridge Green, its cliffs cloaked in birch and holly. A line of fine yews follows their base and they are necklaced, top and bottom, with old, pollarded oaks. There are later plantings, too, of Scots pine and rhododendron, no doubt for the benefit of pheasant shooting, one scourge thus contributing to another. Some trees were thickly clothed in ivy, others hung with the green globes of mistletoe. These surroundings, Dave told me, keep conditions suitably moist for the mosses, and since they love the damp, our British islands, as a whole, are something of a stronghold for them, supporting two thirds of all European species, among them numerous rarities. Shade, too, is important, for their metabolism and their leaves contain a chlorophyll different from that of the 'higher' plants, designed to absorb the particular wavelength of light that filters through a canopy of trees.

The exposed rock face here was up to 33 feet high. Once red or yellow, it had been weathered into browns and greys. Worked on by wind and water, the cliff face was fretted with erosion patterns, lined and streaked or pitted like lace or eaten into to form a rock honeycomb. Here and there, beneath projecting ledges, were rock shelters where Mesolithic families had once sheltered. These varied surfaces, the tops of boulders, the horizontal clefts, the deep shady grooves, were draped in growth; crusts of lichen, dank sheets of liverwort, spreading swathes of moss, and fronds of polypody, hart's tongue and the hay-scented buckler fern. The growth was hung with delicate, looping threads of spider's webbing which, glistening with drops of dew, resembled strings of fairy lights. 'There is an ancient poetry going on between mosses and rocks,' Wall Kimmerer had written, 'poetry to be sure. About light and shadow and the drifts of continents . . . the "dialectic of moss on stone".'

Dave, meanwhile, was uttering little exclamations and sudden whoops of joy and reciting multi-syllabled names too rapidly for me to remember. What did stick, however, were the affectionate descriptions with which he addressed each plant; this one was like a writhing heap of millipedes, this a great tangle of mouse tails. One was a fistful of fingers, another a plantation of young pineapples. One had dreadlocks, while a particularly neat-looking specimen had apparently just been to the barber's. Dave introduced me to the two basic growth forms of the mosses; the acrocarps, which grow upright, often in cushions, with their fruiting capsules held aloft on their stalks, and the pleurocarps which creep across a surface growing in horizontal mats and holding their capsules out from their side shoots. The difference, I was later to learn, would have a significance for the tardigrades.

Mosses are among the most primitive of all the land plants and often defined by what they lack: flowers, fruits, seeds, roots or even a vascular system. It is the absence of these structures that also limits their size but their adaptation is perfect for life in miniature. On our trees, the leaves are held flat and spaced apart, designed to intercept as much light as possible; the leaves of the mosses, on the other hand, are narrow, tight and overlapping, their primary purpose to trap and engineer water. They are the amphibians of the plant world, representing one of the first evolutionary steps of emergence into a terrestrial existence. But, like the amphibians, their emergence is not complete and they have devised all sorts of ways of taking some of that water with them. The shape of the leaves, the way they clasp the stems or curl into tubes, the pitted surface I had seen on the *Bryum* leaves and the long hairs on the *Grimmia*, are all ways of trapping or retaining a watery film, a film that dissolves carbon dioxide and enables it to enter the leaf. But water and weather are fickle and the successful moss needs also to withstand periods of drought. It is, in the jargon, poikilohydric, and when the air around it dries it will stop growing, its cell membranes shrinking and collapsing as it synthesises and stores the enzymes of cell repair for future use. As this happens, it will change colour and shape, becoming less wrinkled or spiralled and crowding in on itself as it survives the loss of up to 98% of its moisture. Here in the moss is a cycle of wet and dry, of dormancy and revival, of life and of waiting for life, a cycle that it shares with the tardigrade and that is at the heart of their relationship. They are partners in their approach to survival.

The day with Dave wore on. The deep bark of a raven reverberated now and then through the canopy while the excitable and high-pitched jackdaws maintained a continuous chatter. It had taken us several hours to advance just a few hundred yards and, though the day was bright, it was also chill and my hands were getting cold. We had looked deep into moss clumps and seen their inhabitants, black springtails and red rove beetles, leading their own lives in their own tiny world. We had found places where layers of moss grew one on top of the other, like overlaid sheets of wallpaper, till, under their own accumulated weight, they were peeling away from the rock surface. We had looked at artichoke galls on the twig tips of yews, at twisted tree roots exposed, clinging and curling on a rock surface, at thin sheets of the rare Tunbridge filmy fern, a fern whose fronds are the depth of just one single cell. Dave, who can be a harsh critic of such things, enthused about the balance between conservation and clearance that had succeeded in keeping the mosses damp whilst not shading them out altogether. 'They are really happy here,' he told me. 'It's the sort of place where delicate things like these can be confident enough to ripple their muscles and say, "In this world, we're big".' The mosses, then, were happy. And if the mosses were happy, Dave was too. And if Dave was happy, then so was I; despite the temperature of my hands.

* * *

I am not too convinced by Dave's metaphor on the 'muscularity' of mosses. But I am sure that they are tough. Even their remarkable adaptability is, however, only amateur stuff when compared with that of the tardigrades, their tenants. Experimentally, tardigrades have been shown to survive conditions of complete desiccation for up to ten years, but this is just one aspect of their amazing resilience. The tardigrade's power of survival against multiple odds is perhaps its most fascinating feature, and certainly its most studied. Stories exemplifying its capabilities abound, though, as with the age estimates for ancient yews, it is a realm where enthusiastic fiction vies with fact . . . a moss sample from Antarctica is said to have been kept in a fridge for thirty years and then produced living tardigrades . . . more living forms are found in a dried piece of moss kept on a museum shelf for a hundred and twenty years. The stories multiply and outreach each other, but the established scientific facts are startling. Tardigrades have survived being subjected to temperatures of −273°C, functionally equivalent to absolute zero, as well as temperatures that, at 150°C, are well above the boiling point of water. They have survived the extremely low atmospheric pressure of a vacuum but also pressures as high as 40,000 kilopascals, six times that at the bottom of the Mariana Trench, the deepest point in the world's oceans. They can survive impact pressures of 900 metres per second and exposure to excessive concentrations of toxins; of carbon monoxide, carbon dioxide, nitrogen, sulphur dioxide and sulphuric acid. At the end of all this they can be brought back to life by the addition of a single drop of water.

It seems rather cruel to have subjected the innocent tardigrade to such experimentation, but it does not quite stop there. On at least three occasions they have also been sent into space. When the European Space Agency launched its Foton-M3 mission in September 2007, they became the first multicellular animals to be exposed to solar radiation and to the vacuum and vacillating temperatures of space, orbiting for ten days at some 260 kilometres above the earth. On their return, 68% of those that had received some measure of protection against high energy radiation survived. Those which had received no protection fared worse, but even here, some were successfully revived and went on to produce viable eggs. Researchers at the Indian Institute of Science made further discoveries in this area in 2020 when they were investigating *Paramacrobiotus*, a new species they had discovered living, rather conveniently, on their own campus. Subjecting it to potentially lethal blasts of ultraviolet radiation, a rather unneighbourly treatment one would have thought, they found that it could survive by forming a protective glowing shield around itself that absorbed the radiation and then emitted it as a harmless blue light. It is also known that other tardigrades, when hit with UV light, glow yellow or orange, though whether this fluorescence has any adaptive significance has yet to be discovered.

Tardigrades, according to a song by the progressive rock band Mute Gods, will inherit the Earth. Theoretical physicists at Oxford University seemed to have adopted a similar view when, in 2017, they looked at the potential impacts of a number of cataclysmic astrophysical events including supernovas, gamma ray bursts, asteroid impacts and the close passage of comets. 'Although human life is somewhat fragile to nearby events,' they concluded, 'the resistance of Ecdysozoa such as Milnesium tardigradum renders global sterilisation an unlikely event'. It is, perhaps, only partially comforting news for the rest of us.

None of this makes the tardigrade what you might properly call an 'extremophile'. It is a word you should avoid should you find yourself in conversation with a tardigradologist. Extremophiles are organisms that live out their lives in extreme conditions, the bacteria, for example, that thrive in sulphur springs or in thermal vents. The tardigrade does not so much 'live' in harsh conditions as endure them. It does so by going into a state of cryptobiosis, a term invented by researchers in 1959 and meaning, literally, 'secret life'. It is an apt phrase for the ability of this, and several other animals, to enter a condition in which, to all intents and purposes, normal life processes cease completely. It has been known about in tardigrades for some time, but what more recent research has revealed is both the complex biochemical processes involved and the fact that these can take several different forms. A tardigrade can go into different types of cryptobiosis depending on the environmental conditions being faced, whether depletion of oxygen levels, increased salinity or freezing temperatures. It is the response to drought, 'anhydrobiosis' or 'life without water', that remains the commonest and if conditions begin to dry out the tardigrade will go through a series of changes that to some extent mirror those in the mosses. To begin with they move together into clumps but then, as conditions become more severe, they begin to shrink, deflating like leaky tyres as they retract their legs and curl in on themselves, their segments folding up on each other, their skin becoming impermeable. Now just a third of their former size, and impervious to almost everything, they adopt a barrel-shape, earning for this phase of their existence the name of a 'tun', from the German *tönnchen-form*. They will now have lost up to 97% of their body moisture and, in what is probably the most remarkable part of the process, replace it by synthesising completely inert sugars, such as trehalose, that prevent the collapse of individual cells and of their organs as a whole. They have, effectively, as Ian Kinchin expressed it, 'turned to glass'.

One of the first to observe the process of tun formation was Lazzaro Spallanzani, the eighteenth-century Catholic priest and naturalist who had come up with the tardigrade name. The process was not described in any detail, however, till 1834, when the Danish naturalist Otto Schulz published an account in the journal *Isis von Oken*. His article describes an experiment in which tardigrades, kept in a completely dry condition for several years, were revived using just a few drops of distilled water.

Alongside his piece the editors ran another; a rebuttal that was longer than Schulz's original article. Professor Christian Ehrenberg said,

> *This report has not convinced me that desiccated animals of no matter what species might revive after death . . . Numerous experiments which I have made with respect to the thoroughly dried infusoria never yielded a positive result . . . I do not believe in marvels beyond those step-by-step processes which can be observed in nature . . . The revival seems to be a deception.*

Perhaps it was that concept of revival 'after death' that caused the Professor so much difficulty. Since so many early naturalists and microscopists were men of the cloth, it is perhaps not surprising that there was a great deal of interest in the question of how one could describe this revival. The boundaries of life and death had been blurred and with Spallanzani describing it as a 'resurrection', the debate became almost theological. It hinged on the still unresolved question of whether some sort of metabolism, albeit at an undetectable rate, continued in the tun or whether the life process had effectively ceased altogether. It is, suggests Paul Bartels, a 'reversible lifelessness', while Ian, in *The Biology of Tardigrades*, had taken a similar view: 'Under these conditions it seems highly improbable that any metabolic activity is maintained . . . it seems probable that there is a total cessation.' If this is the case then it raises a fascinating and almost metaphysical dilemma, one that was set out by John Crowe in an address to the First International Symposium on Tardigrades in 1975:

> *If metabolism actually comes to a halt . . . we are forced into a seemingly insoluble dilemma about the nature of life; if life is defined in terms of metabolism, anhydrobiotic Tardigrades*

must be 'dead', returning to life when appropriate conditions
return. But we know that some of the organisms die whilst in
the anhydrobiotic state in the sense that they do not revive
when conditions appropriate for life are restored. Does this
then mean that they 'died' while they were 'dead'?

* * *

Life and death might not be quite such clear-cut categories as
we had once assumed, and much about the processes that allow
the tardigrade to hover somewhere between the two remains a
mystery. Discovered by a pastor, named by a priest, and raising
questions around the definitions of life and death, the tardigrade seems always to have occupied a realm of knowledge
somewhere between faith and fact. Then, in the late twentieth
century, came a new entry to the theological fray. For the
American creationists, the tardigrade was to become a leading
piece of evidence in their anti-evolutionary case. It was set out
most stridently in 'The "Little Bears" that Evolutionary Theory
Can't Bear!', an article by Joachim Vetter in the 1990 edition of
Creation, the magazine of Creation Ministries International
that claims over 100,000 readers. 'A well-known zoology textbook', runs the opening sentence, 'states that the beginner may
skip the section dealing with Tardigrades'. Innocuous enough
in itself it sets the scene for the later conspiratorial suggestion
that the teaching of tardigrades is deliberately supressed in
schools, lest it lead students to question the truth of evolutionary theory. Tardigrades, we are told, provide 'incontrovertible
evidence that there exist creatures which could never have come
into existence through chance mutation and selection ...
Despite their small size, these "little bears" present some big
problems for evolutionists.'

The 'incontrovertible evidence' is twofold. First comes the
debate over the tardigrade's taxonomic position, an uncertainty
within evolutionary theory being taken as clear evidence of

creationism, as though the whole of biological science consisted of just these two contesting ideas and whatever challenged one were proof of the other. It is a fine example of what the philosophers call a false dualism. A more sustained challenge is to be found in the idea of 'over-adaptedness'. The tardigrade, the argument runs, is adapted to survive environmental conditions beyond even those that would exist after 'a full-scale nuclear war'. Since evolutionists claim that adaptation is 'moulded by chance mutation and selection', the ability of these creatures to withstand conditions that have never actually existed on earth must therefore disprove the theory.

The argument is, in fact, almost an inversion of evolutionary theory. Environmental conditions do not themselves determine genetic variations. They happen at random and are then put to the environmental test. They do not emerge as some kind of fine-tuned act or perfect fit and if a chance mutation throws up something that over-equips an organism, that will do just as nicely, thank you. In the case of the tardigrade, once it has developed a resistance, for example, to low temperatures in the Polar regions, or to the drying out of its mossy home, the fact that it can survive at even lower temperatures or for longer periods of drought, is irrelevant.

Having set out its case, the article returns to its conspiratorial tone: 'Perhaps we can begin to see why it is that students are encouraged to skip over the Tardigrade in their early evolutionary training. After all, one would not want to raise confusing doubts about evolution in the minds of young biologists ...' The idea that the tardigrade is the result of evolution is, in short, 'unsupported by any fact'; it is, in a clever inversion of the usual accusation, 'a position of pure, speculative faith'. But if 'over-adaptedness' is a problem undermining the evolutionary case then how does the creationist explain it? The answer, it would seem, is by their own act of 'pure, speculative faith'. Tardigrades exist because 'it pleased the Creator in his inexhaustible wisdom to call them into existence and to uniquely equip them as He has. They are endowed with capacities . . . which fill us with awe and wonder at the design and planning behind the "things that are made" – even such tiny things.'

* * *

The 'awe and wonder' we could at least agree on, but, though I was making progress in my efforts to get to know the mosses, I had still yet to actually encounter a tardigrade. It called for what I thought of as another expedition, in readiness for which I had already been collecting moss samples. Of course, this would be an expedition unlike any of the others I had undertaken. There would be no big boots and backpack, no exploring of woods or climbing of hills or beating through bracken. It would be rather a matter of staying at home, sitting still and like Leuwenhoek, learning to look at life through a lens. But still it would be an expedition, a journey into the interior, to the terra nova, the world within the world.

We had an old microscope at home. It had been a present to one of our children and hardly used in the intervening years. I dusted it down, replaced the acid-leaking batteries and put it on the kitchen table, held back now only by my lack of experience in using it. It was at this time, and by chance, that my old friend Alan came to stay. Alan and I had been at school together and, in the subsequent years, had shared many a late-night discussion on life issues, as well as on equally knotty problems such as quantum physics, the cosmic anthropic principle, or any other theory we had recently half-digested, our eloquence always increased by whisky. My school encounter with the sciences had ended at O levels. I was, I felt, a victim of that still prevalent prejudice that, at any higher level, the arts and sciences are incompatible and should never be allowed to meet, let alone go out on a date together. It is a belief that has been detrimental to both, and to our general way of looking at the world. Alan, on the other hand, had gone on to start the A level Biology course. Under the mistaken impression that he would therefore demonstrate more competence, I recruited him as the companion on my journey. It was only after we had started that he reminded me he had never actually got as far as sitting the exam.

Together we squeezed out moisture from the moss samples and placed a petri dish beneath the eyepiece. For some time it seemed to be my own eyelashes that I was examining in detail, but gradually I got the hang of it, and there, before me, was the very different world of the miniscule. It felt as if I were entering a new and unfamiliar element, one to which I did not belong but was now a privileged observer. Even the fragments of inorganic matter seemed beautiful in the light of the scope, sparkling like pieces of precious metal.

I would not say our samples were teeming with life, but whilst they were not miniaturised versions of the Serengeti Plain, they soon produced little organisms swimming into view, and each time we encountered a new one, we greeted it with whoops of joy. The first, we found, was a rotifer. It was to this group of animals that the early researchers had ascribed the tardigrades, though personally I could see little to connect them with the thing now in front of me. A transparent, upright tube, it narrowed towards the base to something like a stem, while at the top it widened into a collar. An early name for the rotifers had been the 'wheel animalcules' for around this collar is a ruff of whirling hairs, their constant motion giving the false impression that the whole animal is spinning, in a state resembling the religious trance-like devotion of a dervish. But it is a motion more practical than spiritual, directing food in a stream of water towards the mouth whilst simultaneously maintaining the animal's supply of oxygen.

F.J.W. Plaskitt, who had given such a lively description of the tardigrade, was positively lyrical about the rotifers. They were, he said, 'possibly the most attractive and beautiful group of micro-organisms [whose] lively and sometimes amusing mannerisms have installed them among the first favourites almost as long as they have been observed and examined.' It has to be said that our specimens were among the less colourful of the rotifer's many species, and the comedy of their 'mannerisms' was largely lost on me. But I did find them compelling. One among them seemed to be picking up particles and bits of inorganic matter and arranging them about itself in the petri dish, perhaps to give itself a more amenable surface on which to cling. It seemed to me a real example of the 'thingness of things'. Here was a small and simple organism, so different from me in form and scale, that seemed to be *organising* itself, like I might rearrange the clutter on my desk or a sitting bird rearrange the twigs around it on its nest. It was, as far as I could see, acting with both determination and purpose.

It was harder to ascribe purpose to the nematode that next came into view. A transparent, spindle-shaped, worm-like creature, tapering to a point at one end, it was thrashing and twisting in the water as if frustrated in a continual attempt to tie itself into a knot. Plaskitt, again, had his own description. 'Its movements are peculiar to its type and consist in rather violent curlings and twistings right and left, making little progress thereby, always appearing as if annoyed with its situation or with life in general'. If any organism has reason to be annoyed it is probably not the nematode. It is, perhaps, the most successful form of life on earth. There may be as many as a million different species of them, putting the couple of thousand tardigrades into perspective. Their quantity, too, is staggering. They account for 80% of all animal life on Earth, outnumbering humans by about 60 million to one. Nematodes of one sort or another live on every surface of the planet, leading Nathan Cobb, who studied them extensively, to comment that,

*if all the matter in the universe except the nematodes were
swept away, our world would still be dimly recognizable, and if,
as disembodied spirits, we could then investigate it, we should
find its mountains, hills, vales, rivers, lakes, and oceans represented by a film of nematodes. The location of towns would be
decipherable . . . [and] trees would still stand in ghostly rows.*

Eyepiece to eye with such creatures, our excitement was unsur-
prising and Alan and I worked on through the evening, even
going out into the now dark garden to collect samples of water
from flooded pots or the little leaf-choked pond. Had I access to
Alice's Wonderland cakes I would have eaten one, shrunk down
and swum amongst them, just as people do with dolphins or I
had once with seals. Instead, I called to mind Pastor Goeze and
his 'lions and tigers of the invisible world' and consoled myself
with the thought that down there would be predators and piercing stylets and that this was not my domain. In her wonderful
book *Pilgrim at Tinker Creek*, Annie Dillard regards these
microscopic creatures with an awe that is almost overwhelming.
Describing it as a 'moral exercise', in which the microscope at
her forehead becomes a 'kind of phylactery', she watches the
progress of a rotifer or the flapping convulsions of a round-
worm. They are, she says, 'a constant reminder of the facts of
creation that I would just as soon forget . . . [but] these are real
creatures with real lives, one by one. I can't pretend they are not
there. If I have life, sense, energy, will, so does a rotifer . . . I was
created a clot and set in proud, free motion: so were they . . .
since I've seen it I must somehow deal with it.' I am dealing with
it too. The average size of all living animals, she points out, is
approximately that of a housefly. It means that I, along with the
rest of my kind, will always remain at the lumbering end of
things, suffering the loneliness of the large, experiencing this
other life only through the medium of a metal tube. I was a
peculiar powerless deity, a perpetual observer, peering down

from on high. The one thing my gaze did not alight on, however, was a tardigrade. 'It's hit and miss when you find them,' Ian Kinchin had said to me during our interview, and so far it was entirely 'miss'.

* * *

Contrary to the conspiracy claims of the creationists, my own introduction to tardigrades had come from a Year 7 student returning excitedly from his lessons, and a few weeks after my evening with Alan I had the opportunity to further inflate this irony. I had been invited to lead a walk with pupils at Chigwell School in Essex and, making a pre-visit to the school, I was introduced to the laboratory technicians. I asked if they would be prepared to help me in my search for a tardigrade and to my great relief they agreed.

Returning to the school one bright afternoon in early March, I was escorted to the 'Preparation Room'. It was a long, low room illumined by strip lights and cluttered with cupboards, trolleys, tables and shelves, all of them stacked with equipment. It was a kind of curiosity shop of the sciences where the disciplines shamelessly mingled; Longworth traps jostled with electrical circuit boards, paper cutters shared space with computers, dynamos and batteries balanced next to bowls of sand, and I could only wonder at the purpose of the stirrup pumps. If Dr Frankenstein had run out of equipment mid-experiment, this would have been his first port of call. Posters around the walls instructed me on how to deliver emergency first aid for eyes, or deal with other such contingencies, whilst a sign on the fridge proclaimed that it was for 'Biologically Hazardous Materials Only'. Clearly not a place to keep the sandwiches. An L-shaped table occupied much of the room and at its centre stood eight large bottles of cola and a couple of tubes of mints. They were not, I learnt, the remnants of a science department staff party but equipment for a Year 5 experiment on 'Reversible and

Irreversible Procedures'. It seemed to consist of dropping mints into cola to produce an erupting fountain of foam. To determine the most dramatic effect, the two lab technicians had been experimenting with different brands of cola. For the record, Sainsbury's own brand cola was the most effective of the ten they tried, and also the second cheapest. Nicole and Derek had also been out before my arrival collecting specimens of moss from around the school grounds and they stood in rows of petri dishes around the table, now supplemented by my own.

Since my last efforts with Alan, I had learnt more about which sort of moss was likely to produce the best results. It was not, as I had assumed, the most lush-looking cushions but rather their opposite, the thin, almost powdery forms that had to be scraped from the cracks in tree bark or from the crevices between bricks in a wall. In the words that Dave had taught me, it was the pleurocarps and not the acrocarps that were the most tardigrade-friendly. There was a sound ecological reason for this. A larger, spongier moss would hold more moisture and dry out more slowly but this would also allow it to support larger populations of other organisms, including predators of the tardigrade. It would also house higher concentrations of the fungal spores that pose another threat to tardigrade survival. The thinner mosses dry out more frequently but this diminishes the number of potential predators, whilst the tardigrade survives using its capacity for cryptobiosis. Here was a whole new perspective on the animal. Anhydrobiosis, it seemed, was not just a periodic inconvenience that it had to endure but a positive ecological strategy. The tardigrade welcomed the tun.

Equipped with this information, it was these types of mosses we were now examining, my own specimens having been scraped from the bark of sycamores and sugar maples, from local walls and even from the lock gates of the nearby canal. Even so, and with the use of several microscopes, we were having no luck. We did see more rotifers and nematodes and a beautiful pink mite, and at one time a small piece of mica so convincingly mimicked the shape of a tardigrade that we prematurely began to celebrate. Only its refusal to move over a prolonged period of time convinced us it was inanimate. We were close to giving up when I recalled that Ian Kinchin had discovered his new species not in moss but in the detritus of a gutter. The specimens that Schulz had so memorably revived in his demonstration of cryptobiosis had also been collected from 'gutter sand'. Perhaps this was where we should now be looking. I had hardly formulated the thought before my colleagues were out in the grounds seeking some part of the drainage system that might actually be accessible. The school caretaker, apparently, seemed unfazed by their request for the most congested gutter and directed them to the Arts block where they extracted a sample from above the porch. Thus were the Arts and Sciences materially reunited.

When I saw my first tardigrade, in the gunk of that school gutter, it was a moment as magnificent as any of my other significant encounters with wildlife, whether nesting sea eagles on Mull, wild bison in the Carpathian Mountains or alpine plants on the slopes of Ben Lawers. It was clear as glass, the pale brown contents of its gut clearly visible within. Chubby and segmented, it tapered a little towards one end, its legs stubby and the claws just visible. It crawled like a feeding caterpillar about dark patches of indefinable matter, disappearing frustratingly behind them now and then before reappearing to continue its investigations. At one time it used its back pair of legs to cling on to something whilst rearing up with the rest of its body. My excitement on at last finding the subject of my study was, I am pleased to say, shared. And not just by Nicole and Derek, for several teachers came in to get a view and even interrupted classes so that a string of students from various ongoing lessons were soon forming an excited but orderly queue.

* * *

There was another school, albeit a fictitious one, where tardigrades were the cause of considerable excitement. Though my children loved *South Park*, I had never been a fan of this American cartoon series, but when I learnt that one episode featured tardigrades, a thorough approach to research demanded that I see it. The story centres on the Special Needs Class at South Park Elementary School, where pairs of pupils are working on competing projects in an effort to win the Science Fair prize. Timmy and Jimmy, with a potentially winning approach, have taught water bears to dance to the music of Taylor Swift. Their rivals, Nathan and Mimsy, believing the prize will make them more attractive to girls, have the more everyday idea of making a baking soda volcano. Their only chance, they decide, lies in sabotaging the tardigrade project, by poisoning the creatures with lye. Their attempt backfires, for instead of killing the tardigrades it leads to their 'social advancement' and they begin to dance the hokey-cokey. 'Boys,' says teacher Mr Mackey, 'this might be the single greatest science education project I've ever seen.'

Just as Timmy and Jimmy feel the prize is within their grasp, the 'Men in Black' appear, taking over the school and demanding that every resource now be directed at this single project. 'Water bears are the key to our future,' they exclaim. 'The end is very near and we have very little time.' It transpires, however, that they are not government agents intent on saving the planet. They work for the National Football League where attendances at games have been falling. Their plan is to try and bolster the crowds by turning tardigrades into football fans. And nothing must get in their way . . .

If the plot is improbable, so, too, is the tardigrade; an obvious candidate for both fantasy and science fiction. It has a cameo appearance in the Marvel movie *Ant-Man* and a much more sustained role in the Sentients of Orion, a series of books by Marianne de Pierres. In *Dark Space*, the first of the four, a small group of mercenaries introduce a biologically engineered form of tardigrade to a small mining planet in order to eliminate its inhabitants. The *Tardigrada gigantus*, or 'Saqr', are now huge creatures, freed from their dependence on water, and very, very hungry. Emerging from their cysts they set about devouring the humans, and every other sentient being they can lay their stubby legs on. Their distinctly unpleasant practice is to suck out their juices by penetrating the eye sockets of their prey with their stylets . . . 'Trin's gaze was drawn to the matted clots of darkness in their faces. They seemed untouched apart from the bleeding holes that had been their eyes.'

I had not previously heard of Marianne de Pierres but I have to confess that, mastering the complicated plot, I read on through all four volumes and enjoyed some 1,600 pages. In 'real life' it may be that the possibility of genetic engineering and the tardigrade is going to work in a different direction. Writing in the *Washington Post* in 2016, the science journalist Rachel Feltman reported on research into the unique proteins produced during tardigrade cryptobiosis, speculating on how their ability to protect from radiation exposure might assist with future space travel. 'I'd welcome,' she wrote, the possibility of 'a future of genetically engineered moss-piglet-people populating the cosmos.' Personally, I'd rather face the Saqr.

* * *

The worlds of science and of science fiction come even closer in the idea that tardigrades might have an extra-terrestrial origin, an idea bound up in astrophysics with the theory known as 'panspermia'. It is a theory that is by no means new. The suggestion that life might be universal, that it extends across space and spreads from place to place, has been around in Vedic thought for several millennia. In Western writing it first appeared in the fifth century BC, in the work of the Greek philosopher Anaxagoras. Anaxagoras was an empiricist, a proto-scientist as much as a philosopher, and one of the pre-Socratic group who struggled, through observation and argument, to explain the nature of life and matter. He was the first person to properly account for an eclipse and, in the fragments we still have of his work, expounded the idea that life spreads across the cosmos in the form of 'sper-mata', or seeds, taking root wherever they find the right condi-tions. Across the ensuing centuries the idea of life as universal and migratory, rather than earthbound and static, was revived by a number of thinkers. One of the most respected among them was Svante Arrhenius, a Swedish chemist and physicist who won a Nobel Prize in 1903 for his work on electrolysis. He was also the person who established the existence of the 'greenhouse effect', demonstrating that the accumulation of carbon dioxide in the atmosphere was responsible for increasing global temperatures. Arrhenius believed passionately that life existed throughout the universe and that living microbes were transported through it by the mechanism of solar radiation. His drifting microbes seem strikingly close to the living seed of Anaxagoras, but they were not the only mechanism that had been proposed for the dissemi-nation of interplanetary life. As early as 1864, the German physi-cian Hermann Richter had suggested that meteors might be spreading life from planet to planet, describing a scenario in which a meteor approaching a planet from a glancing angle shed life upon it before bouncing back off into space again. Comets were also coming under consideration and with these larger objects came the suggestion that not just microbes but larger micro-organisms might be carried on or within them, especially the extremo-tolerant tardigrade.

The proponents of panspermia were often people of consid-erable scientific reputation, another among them being Lord Kelvin, the British mathematician after whom the Kelvin temper-ature scale had been named. The theory itself, however, remained on the furthest shores of scientific orthodoxy, regarded, at very best, as an eccentricity. It found a new and vociferous proponent from the 1970s in the work of the British astronomer Fred Hoyle. As well as being a scientist, Sir Fred was well known for his radio appearances and for his writing: science fiction, short stories, radio plays and popular science books. He was also a keen hill walker and achieved the feat of climbing every one of the 283 'Munros', the Scottish hills that are higher than 3,000 feet. During the Second World War he made important improvements in the use of radar and later went on to a distinguished, if controversial, career at Cambridge. He was a founder, then director, of the Institute of Astronomy, his greatest achievement there being his work on 'stellar nucleosynthesis', demonstrating that dying stars create heavy elements that are thrown out into space in their final explosion.

Other of his ideas proved less acceptable to his colleagues. He proposed that the 1918 flu pandemic, mad cow disease and outbreaks of polio were all extra-terrestrial in origin, that flu contagion was linked to sun-spots and, with no experience in palaeontology at all, declared that fossil finds of the prehistoric Archaeopteryx were fakes. He also championed the idea of a steady state universe even when the scientific tide had turned almost totally towards the rival theory of the Big Bang. Ironically, it was Hoyle himself who, in the course of a radio debate, had coined the phrase 'Big Bang'. None of this was received any better than his theory of a cosmic and cometary panspermia. Together with Chandra Wickramasinghe, his research assistant who became for over forty years his co-worker and collaborator, he campaigned against what he called a 'geocentric' view of the universe, suggesting, in *Evolution from Space*, that 'the idea that Earth is the centre of life, the focal point of chemical evolution based on the Carbon atom . . . is as blinkered as the pre-Copernican view that Earth was centre of the Universe.' He was scathing, too, about the idea that life might have arisen spontaneously through the action of random natural processes of a primordial soup: 'Somehow a brew of appropriate chemicals managed to get together ... and somehow the chemicals managed to shuffle themselves into an early primitive life-form'. The chances of such a 'miraculous assembly' of complex biochemicals leading to life on earth he calculated as one in ten to the power of 40,000; a number greater than all the atoms in the universe. It was this statistical improbability that underpinned much of his case. Hoyle and Wickramasinghe were well aware that their theories did not actually explain the origins of life, a criticism that was often levelled against them, but they quoted Lord Kelvin approvingly on the subject: 'The whole universe is so much richer in the opportunities which it affords for solving this fundamental problem, so much richer than the narrow confines of the terrestrial environment, that a theory of life which spans all the universe rather than just one tiny corner of it, has a far better chance of being right.'

By the end of his life Hoyle was an isolated figure at odds with most of his former colleagues. He has been described as a crank, as working on the 'lunatic fringe' and, of course, as a 'maverick'. Perhaps, in this, he is one of a noble company, a company that includes Lynn Margulis and others who by challenging orthodoxy have led to major changes in the way we see the world. Nor was he the only well-known scientist to be propounding strange theories about panspermia. The famous Francis Crick, who won a Nobel Prize in 1962 for his part in deciphering the structure of DNA, also wrote a book called *Life Itself* in which he argued that life had arrived on earth in the head of an unmanned space rocket sent by a higher civilisation as a deliberate act of extra-terrestrial seeding.

Placing these missionary microbes in a protective nose-cone at least escaped the main criticism that was directed at the Hoyle–Wickramasinghe theory – the problem of survivability. Carl Sagan was just one of the many scientists to question the chances of any living organism surviving the violent ejection from their home planet, the prolonged journey across the vastness of open space, subject to vacuum, ultraviolet and X-rays, and finally the high-speed collision with another planet. By the time Hoyle died in 2001, at the age of 86, his propositions about the origin of life on earth had been almost universally rejected and the fringe theory of panspermia ignored or largely forgotten. His colleague, Wickramasinghe, however, never gave up on the cause and was later to return to the fray as one of the authors of a new paper that was to receive widespread media attention.

'Causes of the Cambrian Explosion – Terrestrial or Cosmic?' was published in the journal *Progress in Biophysics and Molecular Biology* in August 2018. As well as Wickramasinghe, it was signed by thirty-two others, including academics from Australia, Belgium, Canada, China, India, Italy, Japan, South Africa, Sri Lanka and the UK, one of them from Hoyle's own erstwhile Institute in Cambridge. It is written in a deliberately unscientific style 'to ensure clear plain-English communications across many scientific disciplines' and in a campaigning tone that is full of sweeping and dramatic assertions.

The argument centres on two big questions surrounding the conventional understanding of the emergence of life on earth: how and when it could have happened. That a hugely complex biochemistry involving the emergence of the nucleic acids of DNA and RNA, and their transformation into life, could occur spontaneously, it describes as a 'superastronomical improbability'. Moreover the timeframe in which this could occur, the period of 800 million years between the stabilisation of the earth's crust and the first signs of life in the fossil record, is held to be far too short. These difficulties led the authors to return to the 'Hoyle–Wickramasinghe thesis of Cometary Biology', stating that, from approximately 4 billion years ago, living organisms such as bacteria and viruses, and sometimes more complex organisms, were seeded on earth by comets, and that such 'deliveries' have not only continued ever since but have helped drive the process of biological evolution.

The paper then sets out to review what it calls the 'avalanche of new evidence', arising from the last sixty years of research, that, the authors believe, provides support for the theory. Starting from Wickramasinghe's own discovery that organic material does exist in interstellar dust, it cites paper after paper dealing with conditions on other planets, evidence of life on Mars, changes in the dating of the earliest life forms, the discovery of microbial material high in the stratosphere, the results of various space missions and the still disputed discovery of microbial fossils in meteorites.

As if rehabilitating panspermia were not a sufficiently daunting task, the article makes a number of other daring assertions. It returns to the suggestion that interplanetary microbes have introduced epidemics and claims comets as the cause of various mass extinctions, including the Cambrian, and for seeding the earth with the life forms that followed. Most strikingly, it claims that the relatively sudden evolution of the intelligent octopus, from the much simpler prehistoric nautilus, is the result of extraterrestrial influences and horizontal gene transfer. Likening the reluctance of the scientific community to accept panspermia to the earlier ridiculing of the theory of continental drift, or the one-time denial that meteorites fell from the sky, the article confidently concludes that, 'all the scientific and societal evidence seemingly points in one direction, an all-pervasive Cosmic Biology, mediated mainly by cometary transfers, being a driver for life on Earth. We believe the signs of this change are now so apparent that one of the biggest back-flips in the history of science is now on our door step.'

Up to this point in the literature of panspermia, the tardigrade has been something of a bit part. 'Causes of the Cambrian Explosion' places it centre stage, beginning with the anonymous, and rather anodyne, quote that heads up the article: 'When presented with the uncanny survival attributes of Tardigrades, a friend exclaimed, "How on earth did they evolve?"' In just a few more pages we get the answer: 'Their catalogue of "living", space-hardy properties is entirely consistent with evolutionary natural selective events acting on Tardigrades evolving in extra-terrestrial space environments. These properties are incompatible with purely terrestrial natural selection conditions either now or 500 million years ago.' The article cites the tardigrade no less than eight times concluding that, 'A plausible evidentiary case for the proof of Cosmic Panspermia could rest on this one example.'

There is an uneasy echo in this sentence of the creationist article by Joachim Vetter. It is not the only time the paper strays disturbingly close to a fundamentalist tract. It has the same strident tone, the same absolute certainty in the truth of its arguments and the same dismissal of other schools of thought. It even contains the same argument on 'over-adaptedness', using it as support for its claims and reprising a very similar line to that of the creationists; that tardigrades 'pose a serious challenge to traditional neo-Darwinian thinking'.

* * *

Its excesses aside, I find the argument of panspermia appealing. The idea that life pervades the universe as a whole seems just an extension of our growing ecological awareness. From the intricate interconnections of a single local habitat through to regulatory cycles and systems that operate on a global scale, it is, surely, only one step further to the concept of a cosmic biome? And how can there not be life out there? Idan Ginsburg, a postdoctoral scholar at the Harvard-Smithsonian Center for Astrophysics, suggests there may be as many as 10 trillion asteroid-sized objects capable of sustaining life in the Milky Way alone. Here, perhaps, is another stage in our growing up. The universe does not revolve around us; the sun does not circle the Earth, humanity is not the end-point of evolution, and our planet is only one among innumerable others. Life, as Brian Cox has commented, 'is far more inevitable than we thought'. And perhaps Anaxagoras was right; it is a universal seed.

If this sounds fanciful, I am not alone in my thinking. Despite some of its eccentricities, panspermia is enjoying something of a resurgence in contemporary science, including researchers to whom the words 'maverick' or 'eccentric' could hardly be applied. Among them is Gary Ruvkun, the award-winning Professor of Genetics at the Harvard Medical School. 'Life on earth got to be super highly evolved very fast,' he argues, 'so fast that its origins must be elsewhere'. The history of Earth offers 4.5 billion years for evolutionary development but the history of the universe as a whole extends that to nearer 14 billion. 'That's why', he says, with a revealing turn of phrase, 'I'm such a religious fanatic about panspermia'. Since 2019 his Ruvkun laboratory has been involved in the SETG project, the 'Search for Extraterrestrial Genomes', developing instruments that can determine DNA sequences on Mars, or any other planetary body.

Another molecular biologist, Pabulo Henrique Rampelotto, testing the resistance of microbes to the environment of outer space, has suggested that, if shielded by two metres of meteorite, a substantial number of spores could survive for up to 25 million years. 'In recent years,' he concludes, 'most of the major barriers against the acceptance of panspermia have been demolished and this theory re-emerges as a promising field of research.' Much of that research, in fact, is already underway, with studies demonstrating, among other things, that meteoric transfer can take place with much less heat than previously expected or that various organisms can survive the impact effects of a planetary crash-landing. And then, in 2020, Japanese scientists revived microbes that had remained dormant on the seabed for as much as 100 million years, raising the possibility that they could also survive the almost interminable transit of space.

None of this provides conclusive evidence of panspermia. It remains a hypothesis, not quite so left-field, perhaps, as it was once regarded, but one that nonetheless remains unproven. And even if it were to be established, would it necessarily explain the origin of the tardigrade? This was the question I had set out to answer and in the end I was not so sure. I had talked to experts, laid siege to the British Library, considered the claims of creationists, explored the mosses and entered the world of microscopy, and at the end of it all there was, it seemed to me, another possible story, and one based on an Earth-bound evolutionary path. It hinged on that relationship with the mosses, and on the existence of a small and little-studied group of tardigrade species that lived out their lives in the sea.

Though tardigrades were first described in 1773, the existence of marine forms was not established until 1851 with the work of French biologist Félix Dujardin, the self-taught son of a watchmaker who went on to become a Professor at the University of Rennes. Though his findings opened up a new area, even today the number of marine species remains low, totalling only 19% of the phylum as a whole. They occur most commonly in the coarse sand of shallow seas, in sub-tidal and inter-tidal regions, and, in quantity, even on breakwaters and beaches. Despite their sometimes fanciful forms they have one significant difference from their land-based counterparts: they do not exhibit cryptobiosis, the ability to suspend all their vital processes and to turn themselves into the tun. The marine environment is far more constant, and the danger of desiccation remote; the marine form has never needed, and never evolved, the same adaptive response. It is a simple but significant fact in the tardigrade story, suggesting that the capacity for cryptobiosis was a later evolutionary development, occurring as some prehistoric form of water bear first left the sea and made that ursine crawl that would lead it onto land. In reality, it was not quite as dramatic as that for there is evidence of intermediate forms and species that would suggest a gradual evolutionary move. Some marine tardigrades, for example, have been revived from beach sand in which they had been stored for several months in dry conditions, indicating an early capacity for an adaptive response. There are also species of tardigrade that live higher up the shoreline, able to cope with the regular short-term drying-out that occurs between tides, while other examples of transitional species can be found amongst the fauna of maritime lichens, those beautiful grey, white and orange crusts that form among the shoreline rocks.

All life begins in the sea and if terrestrial tardigrades did emerge from marine forms, they were mirroring a process that has been repeated over and again in the evolutionary story. Among animals, the amphibians are a well known example, and, as we know, so are the mosses among plants. And it is among the mosses that these emergent tardigrades would find a home. Mosses, like the tardigrade, needed a thin film of water in order to survive and, as I had learnt, every part of their structure is bent to this end. As mosses evolved to respond to desiccation, so did the tardigrade. What is called anhydrobiosis in the animal is poikilohydry in the plant. It seems, then, a reasonable supposition that the two found their way together, the tardigrade evolving from those intermediate forms on the inter-tidal beaches and in the spray-drenched lichens, to withstand the fuller and longer cyclical desiccation in the moss.

There is not, for this idea, the 'incontrovertible evidence' claimed by the creationists, and perhaps not even Wickramasinghe's 'sufficient evidentiary case'. But there is here a probable evolutionary path, and on the principle of Occam's razor, that the simplest explanation is also the most likely, I am happy to go with it for now. Am I disappointed that the tardigrade might not have an extra-terrestrial origin? Perhaps a little, but not much more than that. It seems, to me, just as wonderful a journey, driven by the irrepressible urge for life to extend itself, to have emerged from one element into the great new possibilities of another. There is something inspiring, too, in the thought of co-evolution, of an animal and a plant undertaking this simultaneous journey. And then there are the processes that enable it: the shedding of cellular water, its replacement with complex manufactured sugars, the changing shape, the drawing in, the remarkable formation of the tun. There is so much we still have to learn about this mysterious 'vitrification' and, astral or earthly, the tardigrade fully justifies that description as 'the most miraculous of creatures'. It is, indeed, one of 'the most fascinating animals known to science' and I am grateful to the places it has led my thinking; to an awareness of immensity in both inner and outer space, to questions about the meaning of life and death, and to the concept of a cosmic biome. And I am grateful, too, to both Tom and to his teacher for that first excited introduction to such a fantastical beast.

6

The I in Lichen

Botanists are not necessarily known for their comic side but they do have at least one good joke. It goes like this:

Q: What is a lichen?
A: Anything studied by a lichenologist.

It might not make the cut at the Comedy Store, but it pithily summarises a problem. Here is a ubiquitous, unpretentious and seemingly simple organism that has kept its identity secret. It is almost impossible to classify or define and the scientific efforts to do so have, over several centuries, given rise to both heated debate and deep division. Each stage in the enquiry has led to startling new discoveries that have not only changed the course of science but also impacted on political debate. They have challenged some of the most entrenched and fiercely defended propositions in evolutionary theory and raised such fundamental questions as whether 'species' really exist, or if there is really such a thing in nature as an 'individual'. Here, perhaps, is the real comic theme; that this easily overlooked organism, sometimes no more than a powder on the bark of a tree, has led us to such existential questions as whether I can really speak of an 'I' at all?

It was with all this in mind that I set myself the task of understanding, as best I could, the lichen. It was an organism I loved, for its colours, its shapes and its sheer tenacity, yet I was puzzled why its identity seemed a matter of complete confusion, with our position on it changing frequently and causing such difficulty and division when it did. I wanted to get to our latest understanding of this enigmatic organism, but also to understand the stages in the story through which we had reached it. It was a story, I soon realised, that unfolded like a theatrical drama, in a series of acts, each one with its own intriguing cast of characters.

* * *

There are more than 15,000 known species of lichen, a number that is increasing all the time. Growing on rocks, walls, the bark of trees, bare soil, roofs, fences and even discarded debris, they come in a striking diversity of form. They can be a granular green powder or a spreading tight-clinging crust. They can be shrubby in shape or grow as mats with tiny scales. They can form horizontal 'leafy' expanses, stand upright in strap-shaped fronds or hang downwards in fine hair-like filaments, while some sprout brightly coloured pixy cups in shades of apricot or red. Lichenologists have their own vocabulary for all these forms, names that I found myself wanting to recite out loud like some form of children's skipping rhyme: 'fruticose, foliose; squamulose, crustose; leprose . . .'

Their chemical structure is surprisingly complex, including numerous substances commonly known as 'lichen acids'. Over a thousand of these have now been discovered, most lichen with just two or three of them but some with as many as forty. It is these acids which contribute to the bright colours of some of our lichen, but their exact function remains a puzzle. Lichen do not have flowers or seeds, nor do they have anything as structural as roots, but many do have small peg-like projections on their underside which help them adhere to a surface in a manner that has been likened to Velcro. The physical impacts of these 'rhizinae', when combined with the chemical impact of the acids, means that lichen begin to weather the rocks on which they sit, slowly wearing away and breaking up their surface. Through this, lichen play a vital role in the creation of earth's ecosystem and in the evolution of more complex life forms. They are the makers of soil, bridging the gap between the organic and inorganic worlds as their own death and decomposition adds to the mineral debris they are mining beneath them. They are colonists, the first occupants of bare or newly revealed surfaces and even today, when volcanic islands arise from the sea or glaciers retreat, it is lichen that are the first organisms to establish themselves and to create the conditions in which plants can subsequently take root.

The presence of the lichen acids might also explain why lichen have so few predators. The reindeer of the northern tundra are almost the only exception to this rule, grazing on the great white mats of so-called reindeer moss. At the other end of the scale, with a body only a third the size of its tail, the tiny long-tailed tit is not so much a feeder as a thief, collecting quantities of lichen for its nest. I still have a vivid memory of finding one of these amazing structures when I was a teenager exploring, with my siblings, one of the last stands of woodland still to survive near our home in south-east London. The sign on the fence, its threatening tone reinforced by its bright red margin, read 'Ministry of Defence. Keep Out'. But the fence was invitingly broken here and there and we didn't. Our reward was to come across a long-tailed tit's nest, the size of a decent Easter egg, with its entrance hole towards the top. It was put together with moss and bound with cobwebs, and lined inside with feathers, but the whole of its exterior had been coated with silvery flakes of lichen giving the whole a frosted look. In completing this process, I was to learn, a single long-tailed tit can collect as many as 3,000 lichen fragments. As we examined it gently before retreating back over the boundary fence, it seemed a minor miracle, and certainly not one that the Ministry should be keeping from us.

Unobtrusive though they may individually appear, collectively lichen can frame a landscape. I first became aware of this whilst walking in the Torridon Hills of north-west Scotland. I had climbed with a party to the top of one of the Munros and looked out on a view that seemed to stretch away so far that it eventually dissolved into distance. The surrounding summits seemed rather aloof, each set a little apart from the others like separate citadels, but each a complex of dissected ridges, broken crests and steep, raked slopes. It was a bare and weather-beaten country where the mostly rounded heights seemed to have been rasped and filed into shape by the wind. The sky was clear, the air was clean, and the pale sunshine picked out colours with an intensity that was almost painful. A chief source among them was the lichen. They were orange, lime, lemon, apple green, rust, ivory and grey. They could be soft pastel tones or bright eruptions, or, as the American poet Elizabeth Bishop put it, 'still explosions on the rocks'. Here, where almost nothing else seemed to survive, they were bonding with bare rock, patterning boulders and colouring the hills, surviving in abundance through storm and frost, sub-zero winter nights and unshaded summer sun.

It has been an occasional pastime of mine to fill the odd moments whilst waiting for a train or standing in a checkout queue by devising a personal list of the great natural sights of Britain; ones that could be said to really characterise the country. I would include a bluebell wood in spring, with a density of flowers that seems to have the trees hovering above a ground-hugging mist. There would be the wide skyscapes of an estuary, with mudflats and saltings, and a cold grey sea reflecting an equally grey day, a curlew crying mournfully and a patterned flight of geese overhead. The list would include a sun-pecked autumn beechwood, just as the leaves are turning to glorious gold; the evening shape-shifting of clouds of starlings gathering over a reed bed; the deep purple cladding of heather hillsides still brooding even as they spring into blossom, and, since this is my list and no one else's, there would be an urban view, of the spreading mats of spring violets, transforming neglected front gardens, untended lawns and overlooked grassy fringes. Though everyone is entitled to a list of their own, I have increasingly come to the view that it should always include a landscape of lichen, whether from a mountain scene or from a rocky western shoreline where they coat the tumbled boulders between cliff and sea. Here they mark out the tidal zones in bands of different colour. At the highest spray-splashed point, where the cushions of sea pink and the creeping sea campion form the last bastions of the flowering plant, the lichen are predominantly grey, whether in spreading crusts or the flat-fronded tufts known as sea ivory. At the mid-point they burst brighter, orange or yellow, and, lower, at wave-break, just above the barnacles and brown seaweeds, they give way to an extensive black tar-like coating. In some places this is a subtle gradation and sometimes, I have noted, the grey and the mid-point orange mingle indiscriminately, but elsewhere these zones are as clearly distinguished as the bands of colour on a national flag. Between them, these two habitats, the mountain slope and the salt-encrusted shoreline, illustrate one of the lichen's enigmas: its ability to grow in even the most hostile conditions.

* * *

John Clare, the self-declared 'peasant-poet' of the early nine-teenth century, had an appreciation of the English countryside that encompassed even its tiniest detail. Everything is worth our attention, he declares in his poem 'Shadows of Taste', where,

. . . the man of science and of taste
Sees wealth far richer in the worthless waste
Where bits of lichen and a sprig of moss
With all the raptures of his mind engross . . .

Clare might have seen the richness of lichen but contrary to his assertion the women and men of science, up to this time at least, did not seem particularly engrossed with it. Perhaps this was partly due to their unpretentiousness, compared with the flow-ering plants, but, more significantly, the study of the natural sciences had been almost synonymous with the practice of medicine, and lichen figured little in the armoury of the physi-cian. They do appear in a few of the early Herbals, including the *Grete Herball* of 1526, but in small numbers and in nothing like the great curative panoply of plants. Nor was anyone actually sure what a lichen was; or, perhaps, they were all sure, but sure of something different. They were categorised as algae or mosses or fungi or even as a special class of 'musco-fungi', literally, moss-fungi. Or they were mosses that, checked in their growth, had become degenerate; 'vegetable monstrosities' as they were labelled. Even during Clare's time they were being described as self-generating, induced from decomposing water by the action of warmth and sunlight.

The serious and systematic study of lichen had not really begun until the work of the Swedish country doctor Erik Acharius, born in 1757. Describing Sweden as 'the country of lichens', he set about their dedicated study, collecting, identify-ing and dissecting them. He established the first scientific clas-sification of lichen, putting them in a class of their own which he divided into twenty-eight genera, and the terminology he invented for their detailed structure remains largely in use today. He corresponded widely with fellow enthusiasts, on one occasion dispatching 894 specimens to the Linnean Society in London, of which he had been made an honorary member. Unfortunately, due to customs difficulties, it took them a further two years to arrive. The exchange of specimens was clearly a high point of his practice and might, ironically, have contributed to his death. In August 1819, a colleague by the name of Wahlberg sent him a large consignment of lichen which he had collected in Spain. The gift, according to Wahlberg, was received with such excitement that it caused Acharius to fall ill, and he died within a few days. Another colleague pointed out, more soberly, that the arrival of the specimens, however proximate in time, might not necessarily be connected to the subsequent death. But it made a good story.

It was only a decade later that the German physician, Friedrich Wallroth, was to make an observation about the structure of lichen that was to have a huge significance. The main body of a lichen, thanks to the terminology established by Acharius, is known as a thallus. Upon dissection it reveals a mass of colourless, intertwining filaments. Set within these, in a narrow zone running parallel to the surface, is a band of green cells. It is a small detail, but one that was to provide a great puzzle. What were they and what was their function? As Annie Lorraine Smith put it in her 1921 volume *Lichens*, 'There have been few subjects of botanical investigation that have roused so much speculation and such prolonged controversy as the question of these constituents of the lichen.'

Wallroth himself believed that they were some sort of reproductive structure and named them 'gonidia' from the Greek word *gonos*, meaning 'seed', or 'that which engenders birth'. The same Greek word has given us 'gonads' and even, by mistaken diagnosis, 'gonorrhoea'. Wallroth had noticed the similarity of these gonidia to certain single-celled algae but his belief was that algae, of the sort you can see forming a green scum on the side of a tree or behind a leaking down pipe, represented 'unfortunate brood cells' that had originated from a lichen but were unable to grow and form a thallus of their own. Wallroth's theory of the sexual function of the gonidia was to be the dominant one for the next forty years, until a pronouncement about them that led to a controversial reassessment.

* * *

Grateful for its recent reopening, after two years of pandemic, I had been conducting my researches in the reading rooms of the British Library, summoning arcane texts on the history of lichenology that had to be transported to me from depositories in West Yorkshire that always, in my imagination, figured as subterranean vaults. I had also learnt much from a meeting with Dr Gothamie Weerakoon at the Natural History Museum in London. Both Annie Lorraine Smith, whose book on lichen I had consulted, and her close friend Gulielma Lister of slime mould fame, had once held positions at the Museum, but both of them were 'honorary', or more prosaically 'unpaid', since women were not allowed to hold posts in the civil service until 1919. Now Gothamie seemed to have encompassed both of their work for she bore the splendid title of 'Curator of Lichen and Slime Moulds'. This being the year of lockdown, I hadn't met her, as I would have liked, at the Museum's splendid Victorian buildings, but was constrained once again to an on-screen conversation. I could see her there, in an office deep within the Museum, surrounded by cardboard boxes, shelves loaded with folders and a pin-board with a picture of a prowling cheetah, half-turned as if to check what I was doing. Given that she was wearing a bright floral jacket, I wondered what it was about the flowerless lichens that had first attracted her. It had all happened, it seemed, by fortuitous accident. Brought up in Sri Lanka and completing a Master's in Environmental Sciences, she had been offered the chance to study a group of flowering plants at the University of Arizona. Having young children at the time, she was unable to accept, and when the chance arose for a fieldwork grant, collecting lichen in her native country, she seized it as her only option. And thus she became a lichenologist. 'It was my fate,' she told me with a rueful smile. Her family had been bemused at first, almost hostile, never even having heard of lichen and doubting their serviceability as the basis of a profitable career. 'Will these lichen feed you?' as her father had put it. But she took the job and loved it. 'I was so often away in the forests, coming back with hundreds and hundreds of specimens which were spread out all over the house to dry. They thought I was mad. My hand lens was always round my neck like a permanent necklace.' Later she moved to England with her family and her specimens, which she may well have regarded as family too, training at the Royal Botanical Gardens in Kew before coming to work here at the Museum.

I was enthused by Gothamie's portrayal of those expeditions collecting in the forests and mountains of Sri Lanka, and after long days in the library, determined to get out into the field myself. There was, moreover, a species I was particularly keen to find. In the general silence of the early pharmacopeia on the subject of lichen, there was a notable exception. The first British species to be named in print is the oak or tree lungwort, and it appears in 1568 in the third volume of William Turner's *New Herbal*. Born in Morpeth, in or around 1508, Turner was an ardent Protestant, imprisoned in the reign of Queen Mary and spending several subsequent periods in exile. He nonetheless managed to become an important figure in the history of British botany, and his three-volume herbal was his most important work. I had found a facsimile edition in the British Library and turned to the entry on lungwort. It was printed in a Gothic typeface that I struggled to read, but deciphered to the best of my ability. It ran:

> *The lungwort of the oke drieth and bindeth. It joyneth together and healeth grene wounds and specially them of the lunges. It is also good for outrageous outflowinge of weomens flowes and for spitting of blood and against great lares that endure longe and for the bloodye fire. This herbe is goode for the cough, short windiness and other diseases of the lunges.*

The remedy entailed grinding lungwort together with a mixture of aniseed, fennel seed, liquorice and elecampane, then adding an equal weight in sugar. The result was to be administered as a spoonful each morning and evening. Mixed with salt instead of sugar, it was recommended for 'shortwindiness' in cattle, or for bleeding from the nostrils or mouth in horses.

The lungwort is now a scarce lichen in Britain, restricted largely to sites in the north or west of the country, but my go-to finder in these situations is my friend Dave Bangs and it was he again that I turned to. He recalled that he had once known it from a site in the Sussex Weald, for which reason I found myself, one sunny spring day, driving to a location somewhere south of Tunbridge Wells. It was the sort of morning that makes you want to roll down the car windows and turn some music up loud, but that would have been antisocial, so I'm saying here that I didn't. We met up in a lay-by where my attention was immediately caught by the wealth of lichens on an aged roadside hawthorn. They were a 'nitrophilous' species, Dave explained, ones that could put up with the pollution of passing traffic. He identified a few of them for me but his eye was clearly on the main prize and he hurried me on. We crossed fields dissected by sudden stream valleys, their slopes too steep for modern farm machinery to complete their project of sterilisation. Here, alone, the grass was rich with wild flowers and dotted with dozens, perhaps hundreds, of the humped nests of yellow meadow ants. Looking at one or two of them more closely we could actually see the holes where green woodpeckers had been probing into them for prey.

Some distance further, after we had heard the first skylark of spring closely followed by the distant sound of clay pigeon shooting, we reached the slopes which, Dave told me, were one of the last fragments of ancient deer park in the county. They were thinly wooded with ash, maple, birch, beech and midland hawthorn, with here and there a few fine old oaks. The bracken between them was thick but still low and winter-brown, and paths had been trodden through it by the fallow deer which we would occasionally glimpse in the distance. We made our way to a gill where the trees, some ornamented with mistletoe, grew more thickly, where golden saxifrage flowered on the damper ground and where the sound of chiffchaffs and blackcaps was joined by the continuous trilling of the little brook.

Following Gothamie's example, I was wearing my hand lens 'like a necklace' and we began to examine some of the trees in detail. I can only say that peering through that glass was like entering a different element, leaving the dimension with which I was habituated for one that was new and unknown. It reminded me strongly of the first time I had snorkelled, off the coast of Turkey with my oldest son, and the intensity of entering a place that was totally other, a place where I was a privileged observer but where I could never truly belong. In a similar way I was now not just peering down at the lichen but moving among them and seeing them differently, their contours corrugated, folded, flaked, fluted or ridged, piled like downland hills or bleached coral reefs, their surfaces marked with darkened dots, with moon craters or irregular crowded cups. Dave was pointing out various specimens, explaining that, since we were now in a low pollution zone, this was a very different community from the roadside hawthorn. Few of them had common, English epithets, and the names he was sharing were leaving my head as fast as they were entering, but I remember a *Thelotrema* that looked, under the lens, like goose barnacles massed on the bottom of a boat, an *Usnea* that was a mass of fine filaments, spiked like the frost crystals I have seen hanging from winter pine trees, a *Parmelia* that resembled a bowl of ivory blue corn flakes, and a *Lecanora* that seemed to be covered with tiny treacle tarts. Dave showed me another that looked no more than a dab of white paint, and told me to taste it. I am, in truth, never reluctant to engage with another sense or two, and following his instructions, I moistened my finger, ran it down the lichen, and put it to my mouth. For a few seconds there was nothing, but then a bitter and long-lasting aftertaste that stayed with me for much of the afternoon. It was, I learnt, the standard way of identifying *Pertusaria amara*, once used as a substitute for quinine. The level of detailed study we were now engaged in is known by lichenologists as 'working a tree', an activity that has given rise to another of those botanical jokes. In brief, it involves a party of botanists going on a trip with a lichenologist who never gets further than the car park, but still manages to produce a thirty-page report. That, I think, was the gist of it, though it might have lost something in the telling.

All this was fascinating but it had not produced our Holy Grail. Dave had brought with him a copy of his notes from September 1998 which detailed a 'maple ... with *Lobaria pulmonaria* and *Lobaria virens*, though very shaded on one side by a young beech, the *Lobaria pulmonaria* having disappeared from that side'. We began working our way down the gill, scanning the trees more swiftly, but keeping a particular eye out for the maples. It was inevitable, after some time spent fruitlessly in this exercise, to become somewhat pessimistic and I began focussing on the bird song instead. A stock dove was cooing deeply, like a child with whooping cough, and the high clear call of a nuthatch sounded overhead. Dave was keen to see it and we followed the sound till we located it high in the branches of a tree in the gulley. Suddenly Dave was shouting at me, in a most unbotanical manner: 'We've found it, we've only bloody found it.' It was not so much the nuthatch he was referring to as the fact that, searching for it through our binoculars, we could see a rich growth of the lungwort. It was not on a maple, nor even an oak, but a fine old ash, thick with growths of mosses, liverworts and lichens, almost tropical in their luxuriance. Among them the tree lungwort was clothing one side of a bough, lobed and leafy and projecting like miniature staghorn ferns, grey-green to pale brown in colour but a brighter clear green where it was dampened by the adjacent moss. It did, loosely, approximate to the shape of a lung. One of the prevailing schools of herbalism in Turner's time was the Doctrine of Signatures. It taught that God had given every plant a sign in its shape or colour which determined how it should be used. It was from this supposed pulmonary resemblance that Turner's medical recommendations had arisen.

Dave was now conducting a thorough but unsuccessful examination of the surrounding area for further specimens. Not normally of a mystical bent he became convinced that the nuthatch had led us deliberately to the spot. I made no comment but I did notice that the first wood sorrel of spring was flowering beneath it. An old country name for the wood sorrel, I remembered, was the 'alleluia plant'.

* * *

On 10 September 1867, at the Annual General Meeting of the Swiss Natural History Society, Simon Schwendener, the newly elected Professor of Botany at the University of Basle, rose to present a paper. Born in 1829 in Buchs, a small village in the Swiss canton of St Gallen, he was the only son of a farmer and had therefore been expected to take over the family farm. He was also, however, a brilliant student with a growing interest in science. When his family's financial circumstances prevented him from going on to higher education, he worked instead as a teacher at the village school while taking science classes in his spare time. A small inheritance, combined with the ability to live on almost nothing, eventually allowed him to get to Zurich University where he began an intensive study of lichen. As his academic career progressed, so did his skill with the microscope, leading eventually to the appointment at Basle. In his presentation to the conference, he put forward, for the first time in public, his striking new idea about the lichen gonidia. They were not, he suggested, reproductive structures at all; in fact the gonidia and the layer in which they were to be found were not even related. The fibrous layer was a fungus, the gonidia living within it were an alga. The lichen, in other words, was not a single organism at all, it was two.

The President of the Natural History Society was politely sceptical of the paper; elsewhere it was greeted with outrage. Here was a suggestion that two separate living things were working together, the alga with its chlorophyll processing food for the organism as a whole. It was a suggestion that defied common sense and violated the orthodoxy that all living things were unitary and autonomous. It certainly didn't fit with the now well-established Linnean system of classification, with its defining and labelling of distinct species, and therefore the taxonomists were against it. And so, too, were the majority of contemporary lichenologists. The new 'Schwendenarian hypothesis' was, suggested Lauder Lindsay, 'merely the most recent insistence of German transcendentalism, applied to lichen'. American newspaper editor and lichenologist Henry Willey asserted that it was a theory to which 'no true lichenologist will be able to assent', the botanist Mordecai Cubitt Cooke was confident that 'even if endorsed by the 19th century, it will certainly be forgotten by the 20th', and the *Lichen Flora* of 1879 dismissed it as 'purely imaginary . . . the baseless fabric of a vision'. These were but mild rebukes, however, compared with the vitriol that was to pour from the pen of William Nylander, the leading lichenologist of the day. Born in Finland, but now living as a recluse in Paris, Nylander had devoted his whole life to the study of lichen producing a constant stream of books and papers, most of them in Latin. According to G.C. Ainsworth, in his *Introduction to the History of Mycology*, Nylander 'never admitted to any criticism of his methods, his opinions once stated were never revised . . . [He] regarded as personal enemies those who dared to differ from him.' And just such a one was Schwendener. His work was 'absurd tales' or *stultitia Schwenderiana*, the product of 'Schwendener the Simpleton'. Schwendener's ideas, he wrote 'scarcely deserve even to be reviewed or castigated so puerile are they – the offspring of inexperience and of a light imagination.'

The most important British lichenologist of the day was the Rev. James Morrison Crombie, a lecturer at St Mary's Hospital, London. Where Nylander led, Crombie followed, ridiculing this 'sensational Romance of lichenology' with its 'unnatural union between captive algal damsel and tyrant fungal monster'. Nylander, he asserted, had convincingly reduced the Schwendenarian hypothesis 'to the nothingness from which it ought never to have occurred.' Many years later, and long after his death, the righteousness of this man of God was itself called into question. When his herbarium collection at the Natural History Museum was re-examined many of his most spectacu-lar discoveries of 'new' British lichen were found to have been deliberately mislabelled and had in fact been collected from abroad. This deceitful attempt to boost his reputation is over-looked in most histories of lichenology and Oliver Gilbert, who does deal with it, is forgiving of what he calls 'these lapses', preferring to remember Crombie for his 'outstanding and enlightened contribution to lichenology'. It is a magnanimity of which Crombie himself seemed incapable.

A surprising bit-part player in this unfolding drama was Beatrix Potter. Before her better-known work as a children's author she was a proficient amateur mycologist, producing a series of 350 highly accomplished illustrations of fungi and mosses. She had begun this work on family holidays in Dalguise, encouraged by Charles McIntosh, the local postman. Gradually her interest had become more scientific and she had begun to dissect her specimens and include cross sections of details like the fungal gills. Working with a microscope, she also began to illustrate their spores, and to experiment with germinating them on glass plates. She was obviously aware of the controversy then raging and, writing in her journal in December 1896, describes how she had taken what she thought was a lichen to show George Murray, the 'Keeper of Botany' at the London Natural History Museum. He told her that it was actually a fungus that resembled a lichen. This had led her to ask for his views on the Schwendenarian hypothesis. 'He was', she wrote, 'very high-handedly contemptuous of old-fashioned lichenologists'.

Potter wrote her journals in code and they remained undeciphered until 1966 when they were translated in a volume by Leslie Linder. 'It sounds', wrote Linder in a footnote to this entry, 'as if Mr. Murray was casting doubt on the possibility of two partners living in symbiosis . . . whereas BP was apparently convinced of this.' This was to become the standard view, with Potter cast as the female amateur scientist carrying out work to prove Schwendener's theory and being rebuffed by the closed ranks of a patriarchal Victorian establishment. In fact the contentious line can be more easily interpreted in the opposite way, and this has been the direction of more modern scholarship. It was Potter, it now seems, who was defending the 'old-fashioned lichenologists', the ones she had been brought up on like Nylander and Crombie, and it was Murray who was convinced of the dual hypothesis. Within a few years Potter had turned from mycological illustration to children's books. If she was on the wrong side in the lichen controversy she has at least the consolation that *The Tale of Peter Rabbit* has now sold some 250 million copies.

George Murray was just one of an increasing number of scientists who, despite the critical onslaught, were becoming convinced of the Schwendenarian case. It would be, however, over seventy years before Eugen A. Thomas successfully cultured a lichen in his laboratory by combining its two known components. The experimental proof that the dual hypothesis had so long awaited, had finally arrived. Schwendener, meanwhile, continued to devote his life to botanical investigations, studying how the principles used in mechanical construction might also apply to the structure of plants. By the end of his working life he was a Professor in Berlin, a patient and amiable but rather lonely man, though surrounded by a close-knit group of students noted for their camaraderie and their devotion to their work. In the privacy of his Berlin home he wrote poetry, mostly in praise of nature but with a few rather shy and tender love poems referring to a sensitive man who, because of his economic station, could not dare to hope for lasting happiness. He never married and on his death in 1919 left all of his possessions to the primary school in Buchs. Recalling a career of more than fifty years he had once said of himself, 'Fighting for science, I have grown old, but I have succeeded'. The full implications of his lichen-based theory, however, were yet to come.

* * *

There was, meanwhile, a completely different matter involving lichen that had been the subject of intense, if less heated, debate and which had caught my attention. It concerned the story of the 'manna', a story that had been important enough to figure in the scriptures of three faiths. The more literally minded scholars made considerable efforts to determine what, exactly, it might be and among the contending explanations was the suggestion that this heavenly food supply might have been a lichen.

The story of the manna appears twice in the Bible, once in the Book of Exodus and again in the Book of Numbers. Led by Moses, the tribe of Israelites has escaped from enslavement in Egypt and crossed the Red Sea, which miraculously opened before them. The following years are ones of nomadic wandering in the 'Wilderness of Sin', the area we now know as the Sinai Peninsula. It is not long before the people are complaining about their conditions and looking back nostalgically on their time in captivity, when 'we sat by the meat pots and ate bread to the full'. Once you are free of them, it is easy to romanticise past tyrannies, just as some Russians today look back with longing on the iron certainties of Stalin. God responds to the people's complaints with a certain amount of forgivable grumpiness, but nonetheless lays on some remarkable provision. Every evening quails appear and 'cover the camp', while each morning a heavy dew evaporates to reveal 'a fine flaky substance, as fine as frost' lying on the ground. The Israelites call it 'manna', meaning, literally, 'What is it?' Moses replies that it is 'the bread which the Lord has given you to eat'. Each morning they are to gather it, grind or pulverise it, and bake it into cakes or loaves, whose taste is like 'fresh oil' or 'wafers made with honey'. In a lesson in sustainable consumption, they are instructed to gather only as much as they need for the day ahead; anything left over at the end of it will 'gather worms and stink', the one exception being the day before the Sabbath, on which they should collect enough for the coming two days.

There is, in the story, something more than a miraculous emergency meal service and its interpreters have given it several layers of meaning. Moses himself hints at its lasting importance when he instructs that a sealed pot of manna be put aside to serve as an instruction for generations still to come. As the 'bread of heaven' it appears again in the writings of the prophet Nehemiah and also in several of the Psalms, described, among other things, as 'the food of angels'. In Christianity it makes its way into the New Testament, where Jesus in person becomes the 'bread of heaven', a sort of permanent replacement of the transient manna. It is essentially, therefore, the bread of the Christian mass, and it appears again when Jesus is teaching his followers how to pray, setting out for them what has become known as the 'Lord's Prayer'. The request to 'Give us this day our daily bread' echoes the morning appearance of the manna and the injunction against excessive acquisition. And finally it appears in the Quran, as a symbol of God's benevolence and mankind's ungrateful response: 'And We gave you the shade of clouds and sent down to you Manna and quails, saying: "Eat of the good things We have provided for you:" [But they rebelled]; to us they did no harm, but they harmed their own souls.'

Alternative theories on the nature of this manna have been circulating for a very long time. As early as 1483, Breitenbach, the Dean of Mainz, travelling in the Middle East, had noticed thickets of tamarisk trees growing in some of the wadis. From these dripped a sweet and sticky substance. 'In every valley throughout the whole region of Mount Sinai', he wrote, 'there can still be found Bread of Heaven, which the monks and the Arabs gather, preserve and sell to pilgrims and strangers who pass that way. [It] falls about daybreak like dew or hoarfrost and hangs in beads on grass, stones and twigs. It is sweet like honey and sticks to the teeth.'

It clearly appealed to Breitenbach's sweet tooth for he bought a lot of it. Only long after his time was it realised that the substance was not directly from the tree itself but from the plant-lice that fed upon it, sucking the sap and excreting the excess sugars, much in the same way that aphids drop honeydew on an urban pavement. As an explanation for manna, it was only one of several involving insects. According to an article in the *Independent* in April 1996, 'What the Israelites were gathering was the cocoon of the parasitic beetle Trehala manna . . . found on thorn bushes in the Middle East'. The substance they produced was described as 'a white crystalline carbohydrate' and was named as trehalose. Another theory appeared in 2010 in the academic journal *Opticon 1826*, where Roger Wotton postulated that manna could have been composed of the compacted bodies of swarms of non-biting midges, blown some distance from the water bodies where they congregate. Such midges are collected in regions of Africa, he pointed out, where they are dried, ground to a flour, then mixed with water to produce what are known as 'kungu cakes'.

An even stranger explanation was that put forward by Immanuel Velikovsky in 1950 in his book *Worlds in Collision*. The book had been an instant bestseller and its arguments caused a sensation. In what became known as the 'Velikovsky affair' it also united most of the academic world in opposition. Velikovsky's case, based largely on the interpretation of ancient texts, was that Venus had originally entered the solar system as a comet. Passing dangerously close to the Earth before settling in its current planetary orbit, it had led to a sequence of cataclysmic events, recorded in the myths and writings of many earlier peoples, including the Hebrew scriptures. These terrifying global impacts, including erupting volcanoes, earthquakes, tsunamis, inversions of the magnetic pole and great clouds of dust and debris, accounted, among other things, for the great plagues of Egypt, the parting of the Red Sea and the 'cloud by day' and 'pillar of fire' by night, said to have followed the Israelites throughout their wilderness wanderings. The volcanic and cometary dust, according to Velikovsky, 'saturated the atmosphere with floating particles'. Overloaded with these compounds of carbon and hydrogen, it discharged them from the clouds in the same way that rain or hail is formed. It was these falling compounds that constituted manna.

Despite his self-characterisation as a 'supressed genius', Velikovsky's ideas were comprehensively dismissed, both by scientists who questioned his grasp of orbital mechanics, and by historians who rejected his highly selective use of sources. The insect-based interpretations of the origin of manna are hardly more convincing. The two biblical accounts are quite specific about its physical appearance, likening it to coriander seed and describing it as white or yellowish in colour. There is no resemblance here to a compressed mass of midges, nor is there any mention of trees or bushes from which beetle cocoons or the suckings of plant lice might have fallen. Neither would such seasonal occurrences have provided either the continuity or the quantity needed to feed a whole tribe over a protracted period of time. But this is precisely where the lichen explanation comes in.

Lecanora esculenta is the 'desert lichen', abundant in North Africa and West Asia where it grows on rocks and soils. In one wadi, in eastern Libya, it was once reported as covering an area of 2,100 square kilometres. Easily broken off and blown into heaps it has been found cloaking the ground to a depth of up to twenty centimetres. Its irregular contorted lumps vary in size from a pea to a small nut, pale brown or whitish in appearance with a white interior that has the consistency of flour. It is said to have a slight but pleasant sweetness and, once prepared, to keep well in a sealed container. Significant 'falls' have been reported as far apart as Algeria and Iran, and, while leading a French expedition in the Sahara in 1867, a General Yussuf decided to try out its value on his soldiers. Made without meal, he said, it was friable and lacked consistency. Mixed with just one tenth of meal by weight, it was similar in taste as well as texture and consistency to the soldier's ordinary bread. Perhaps Moses, who was brought up as an Egyptian, had seen its usage there and was now adapting it to the needs of his people. The lichen theory is not without critics, but for anyone, like the Israelites themselves, seeking an answer to the question 'What is it?', *Lecanora esculenta* is the best that we have.

* * *

The storm of criticism faced by Schwendener's dual hypothesis owed one thing at least to the way he had presented it. The relationship between fungus and alga in the lichen he had characterised as that between master and slave. The fungus was a parasite, 'though with the wisdom of statesmen' the algae were 'its slaves which it had sought out and pressed into service'. He named them 'helotes', after the class of slaves in ancient Sparta. It was an approach that deterred even some of his erstwhile supporters and as more and more examples of symbiotic relationships were uncovered over the next few years they began to be considered in a very different light. As early as 1873, while conducting his own research into the crustose forms of lichen, Albert Bernhard Frank declared that 'the association between lichen hyphae and the gonidia, is something more than simple parasitism'. Far from being hierarchical, with dominant and subservient roles, the arrangement was one of mutual benefit, with the alga using its chlorophyll to produce food, and the fungus supplying essential minerals as well as a protective environment. A lichen is essentially no more than a collective arrangement between the two. To describe this arrangement Frank coined the cumbersome term 'symbiotismus'. At about the same time, the German surgeon Heinrich Anton de Bary, who had already foreshadowed Schwendener's work on the dual hypothesis, was also developing his thinking on 'the living together of dissimilar organisms'. It was his simpler term of 'symbiosis' that was to catch on. The word has had a confusing history. Technically it is an umbrella term covering all forms of 'living-together', including parasitism, but in common parlance, and often in scientific usage too, it has come to refer specifically to those relationships which are co-operative and of mutual benefit.

Within a few years of the Basle paper, other scientists, many of whom had known or worked with Schwendener, were finding further examples of symbiosis; in sea anemones, in fungi and in the root nodules of beans. As the years passed, more and more were to come to light, revealing such relationships in corals, sponges, deep sea slugs, aphids and termites, and in the sort of story that might figure in a modern nature documentary – the ox-pecker feeding on the back of a rhinoceros or the little wrasse fish cleaning the teeth of sharks. Despite all this, symbiosis continued to be seen as something of a biological curiosity, an evolutionary sideline, fascinating, perhaps, but not part of the mainstream of science. By the beginning of the twentieth century, however, there were new and radical voices seeking to give it a much more central role. It began in Russia where a number of biologists were promoting the idea that symbiosis had actually played a key role in the evolution of life on earth. Their papers were never translated into English and, as far as the West was concerned, their voices went unheard. Within a few years, however, a number of European and American scientists were independently coming to a similar conclusion. 'All living things,' claimed Paul Poitier in 1918 at the Monaco Oceanographic Institute, 'all animals from amoeba to man, and all plants, are constituted by an association of two different beings.' Nine years later, Ivan Wallace, who had been looking at symbiotic relationships at the microbial level, suggested that the establishment of such intimate 'microsymbiotic' complexes was 'the cardinal principle involved in the origin of species'. Another such researcher was Félix d'Hérelle who discovered 'phages', bacteria which also contain a virus and which, during the First World War, were widely and successfully used as a medical treatment. Significantly, he labelled them 'microlichens' and came to the controversial conclusion that 'symbiosis is in a large measure responsible for evolution'.

These were, however, lone and uncoordinated voices in a hostile environment. Such ideas, proclaimed the famed cell biologist Edmund Beecher Wilson in 1925, were 'too fantastic for present mention in polite botanical society'. It took until the

1960s and the work of a remarkably resilient woman to turn such voices into a coherent theory. Lynn Margulis was born into a Jewish Zionist family in Chicago in 1938. She was an agnostic and an evolutionist, a supporter of Darwin who rejected neo-Darwinism. She was also a woman who was used to, and perhaps thrived on, opposition. Rebellious, contemptuous of dogma and firm in her convictions, she once pronounced, 'I don't consider my ideas controversial. I consider them right.' One of her early grant applications had been returned with the words, 'Your research is crap. Don't ever bother to apply again'. Her seminal paper, 'On the Origin of Mitosing Cells', written in 1967 while she was still a young faculty member at Boston University, appeared in the *Journal of Theoretical Biology* only after it had been previously rejected fifteen times. In it she argued that some of the most important moments in evolution had arisen from the coming together of different organisms, the most significant step of all being the development of the complex cell on which all subsequent life forms were based. For untold millennia life on earth had consisted of simple, single-celled organisms like the cyanobacteria that floated changeless on the surface of vast soupy seas. In a step described by biologist Ernst Mayr as 'perhaps the most important and dramatic event in the history of life', somehow the complex cell had arisen with a membrane, nucleus and 'organelles'. It was the beginning of an evolutionary explosion and it had happened, according to Margulis, because one or more forms of bacteria had entered another and set up home there, establishing a relationship of mutual benefit. From this came the startling implication that every single cell in a plant and in an animal, ourselves included, is the embodiment of this fusion. Every cell in our body represents the synthesis of different beings. We are all, it seems, macrolichen.

Such an outrageous idea was predictably greeted with outrage. Across years of fierce criticism Margulis became famous for her tenacity, until, in 1978, the work of Robert Schwartz and

Margaret Dayhoff, produced powerful genetic evidence in her support. The DNA of the nucleus, they demonstrated, was different from that of the rest of the cell. Within another two years further research had shown that the mitochondria in animals, and the chloroplasts in plants, are also composed of significantly different genetic material from the rest of the cell. As her ideas became more and more widely accepted it became clear that symbiosis was far more than an intriguing but incidental side alley. From the initial study of gonidia in lichen had come a principle shown to play a pivotal role in the story of life on earth.

Margulis was perhaps a natural iconoclast. Or perhaps it was because of her earlier experiences of rejection, that she came to identify with other scientific mavericks. For some time she co-operated with James Lovelock in the development of the Gaia hypothesis, a notion that is almost a global version of symbiosis in itself, suggesting that the biosphere as a whole is a single self-maintaining adaptive system. Ironically, this discoverer of a fundamental biological truth went on to support more and more outlandish theories, her work becoming increasingly controversial at the very time that her earlier ideas were gaining acceptance. She promoted a paper claiming that larvae and the adult insects that emerge from them have evolved from different ancestors. She was branded an HIV denier for claiming that the illness was actually a form of syphilis, and went on to claim that the 9/11 attacks on the World Trade Center in New York had been a staged operation. The three buildings had collapsed, she argued, as the result of a deliberate act of controlled demolition set up to justify wars in Afghanistan and Iraq and to mount an attack on civil liberties. It is something of an irony that it was her first husband, the biologist Carl Sagan, who had led the establishment charge against eccentric ideas in the earlier affair of Immanuel Velikovsky.

Given its enormous implications, it was not surprising that the idea of a symbiotic origin for the complex cell had received such a hostile reception. For one thing, bacteria had previously been seen as pathogens rather than partners, organisms that were associated with death and disease and therefore almost universally characterised as the 'enemy'. But there was something much more than this. The new theory of 'endosymbiosis' was posing a threat to the entrenched scientific orthodoxy of the time. Darwin's original theory of evolution had been combined with the emerging understanding of genetics to produce what was known as the 'modern synthesis' and like any orthodoxy it was adhered to with a doctrinal and devotional fervour. 'We are dealing', Margulis put it, 'with clashes between religious doctrine and its reformation'. Endosymbiosis posed a number of specific challenges. Evolution, said the conventional doctrine, was essentially gradual and the result of random genetic mutations. Here now was a proposal that it could progress in 'sudden' leaps and bounds. The idea that characteristics could be acquired in the course of a lifetime, and then passed on, was also heretical. It had once been proposed by Lamarck who had been roundly derided for it, and now it was back on the table. Most of all, evolution had been seen entirely in terms of a contest for life. It was harsh and unforgiving with new mutations tested in the crucible of competition, and here was the suggestion that co-operation, as much as competition, was one of the great engines in the emergence of new species.

The science fiction stories I read as a youth sometimes contained the conceit that all the world's troubles could be solved if only we had a global government of scientists. Free from nationalistic, religious and political prejudices, they would constitute a totally rational and objective team, dedicated to the earth's advancement. The history of evolutionary theory demonstrates the opposite, that rather than standing in a rarefied realm of its own, science is integrally intertwined with the thought and politics of its time. Just like every other aspect of society, it is shaped through the prevailing patterns of its day. Darwin's insistence that adaptation, and the formation of species, could be explained by conflict and competition was very much within the current of the mid-nineteenth century. *The Origin of Species* had been written in 1859, during an era of aggressive Victorian capitalism. Among his acknowledged influences were Adam Smith, with his theory of the division of labour, and Thomas Malthus, whose *Essay on the Principle of Population* argued that unchecked arithmetic population growth would intensify the struggle for existence, on which basis he opposed the poor laws and public charity. It was in this current that Darwin had subtitled his book *The Preservation of Favoured Races in the Struggle for Life*, that Tennyson wrote of 'nature red in tooth and claw' and that Thomas Huxley, Darwin's most vociferous champion, had spoken of life as 'a gladiator's show whereby the strongest, the swiftest and the cunningest survive to fight another day.' The philosopher Herbert Spencer, the most famous European intellectual of his day, took Darwin's ideas very firmly into the human sphere promoting 'social Darwinism' and using it to champion laissez-faire social policies. It was Spencer who was to coin the phrase 'survival of the fittest'. This ruthlessly competitive conception of evolution was perhaps to reach its eventual peak with Richard Dawkins' book *The Selfish Gene*, the antithesis of Lovelock's 'Gaia' and written just three years before Margaret Thatcher came to power. As the great Stephen Jay Gould put it in *The Panda's Thumb*, 'On issues so fundamental as a general philosophy of change, science and society usually work hand in hand.'

Margulis herself was clear about the political relevance of her evolutionary theories, once denouncing the proponents of 'orthodox neo-Darwinism [as they] wallow in their zoological, capitalistic, competitive, cost-benefit interpretation'. It was not surprising, in the context of the Cold War, that evolutionary ideas incorporating co-operation were seen as not just wrong, but virtually communist in tone.

* * *

A further challenge in these new ideas was to the existing concept of the biological 'family tree'. That tree, it seemed, was being unceremoniously felled. In the classical theory, everything proceeded from a single source, one thing inexorably giving rise to another, with variations spreading out from a central trunk into boughs and branches and twigs. It was a design of constant progression, with humanity often shown at its very summit. Endosymbiosis suggested otherwise, and something far more messy; that life, and seemingly similar life forms, could emerge more than once, evolving separately at different times, in different places and from different ancestors. It was less of a tree and more of tangled bush. The lichen was, again, the great exemplar. Acharius, Wallroth, Nylander and all those earlier lichenologists had struggled to place lichen somewhere in the linearity of life; to find, in essence, its single common ancestor. But there was no such thing. Lichen, it now appeared, had arisen not just once but many times over, every time a new association was forged. They were more of a lifestyle than a taxonomic group.

The fact that this lifestyle has recurred so often proved just what an effective partnership it was. 'The numerical significance and the enormous ecological success of the lichenised fungi in the most diverse terrestrial ecosystems', wrote Margulis, 'impressively illustrates the innovative force of the symbiotic way of life'. Lichen have been able to cover as much as 8% of the earth's surface, an area greater than that occupied by the tropical rain forest. They occur on every continent and in even the most extreme environments, from arctic tundra to the driest desert. The common lichen *Xanthoria candelaria* has been found in Antarctica, still photosynthesising at temperatures as low as $-16.5°C$. Lichen can grow not just on bare rock but within it, insinuating their growth between its grains. Some survive on no permanent surface at all, taking the form of perpetual, rootless nomads, or a terrestrial version of the jellyfish, blowing here and there as the wind dictates. Some flourish on toxic slag heaps and all are resistant to radiation. They have even been sent into space and exposed to wildly fluctuating temperatures, cosmic radiation and the full spectrum of solar light. Their survival rate was 100%. They are, in other words, examples of the true extremophile. They are also among the slowest growing things on earth, some expanding radially by only as much as 0.09 millimetres a year. But in that slowness is their longevity, for the oldest living organism on earth is neither a yew nor a bristlecone pine, but a lichen, found in northern Sweden and dated, still going, at 8,600 years.

Of all its startling qualities it was, no doubt, the lichen's long, slow lifespan that suggested it to John Wyndham as the subject for a novel. I was, throughout my teenage years, a huge fan of Wyndham, reading just about everything he wrote; not just the best-known novels like *The Day of the Triffids* or *The Midwich Cuckoos*, both of which were made into films, but also lesser-known works like *Chocky* and *The Seeds of Time*. My unqualified favourite was *The Chrysalids*. It was set in a world that was post-nuclear holocaust and reduced to fundamentalist, rural communities, ruthlessly rooting out any mutant forms arising from the residual radiation. The story centres on a group of children and young people who must conceal a new mutation – that they can communicate telepathically. Brought up myself in a fundamentalist family, and harbouring my own dreadful secret, that I no longer believed, my identification with them was absolute.

There is a ring of truth in the assertion by Brian Aldiss that Wyndham's work is composed of 'cosy catastrophes', but nonetheless I loved it. When I heard, in 1960, that his latest book had come out, I begged my parents to buy it for me as a Christmas present. It was, of course, in the time before online ordering and they must have traipsed some way to find it, for bookshops were not a feature in Bermondsey. Nonetheless, there in my stocking on Christmas morning was *Trouble with Lichen* and I must have finished it before the end of Boxing Day. Some sixty years later I no longer have that copy, nor could I recall a word of its contents. I was determined, therefore, to procure and re-read it.

It is not, it turned out, among his better works, which might explain my amnesia. Its rather clunkily written story concerns two biochemists, one male and one female, who discover that a certain rare lichen has the power to slow down the aging process, extending human life by at least another 200 years. Working separately, they both try to extract the active ingredient. The plot does at least pose the interesting question of how society would respond to such a discovery and follows the different responses of the two. Francis Saxover is concerned that making public the life-prolonging powers of the lichen would lead to corruption and chaos – and contribute to world over-population. His colleague, Diana Brackley, by contrast, believes that a longer lifespan would give people both the time and the incentive to address social problems, providing the antidote to our tendency for short-term perspectives. In her analysis, the problem lies not with individuals but with the institutions, political, administrative and commercial, that would, she argues, try to control and suppress such a life-extending drug.

It is with these issues, and with Wyndham's uneasy efforts to adopt a feminist perspective, that the book wrestles. There are, of course, plot twists along the way and while I do not wish to provide a spoiler, I will confide that Diana and Francis somehow resolve their different perspectives and, having both been previously injected with the lichen extract, set up house together to live happily almost ever after. The lichen being rare and only found in China, their goal is to produce a synthetic version thus meeting her demands whilst dealing with the scarcity issue that worries him. After all, as Diana pronounces, 'we can't both sit down here and do nothing for two or three hundred years, can we?'

* * *

By the early twenty-first century, with its dual nature securely established, it might reasonably be assumed that the riddle of the lichen had been finally unravelled. In fact there was much more to come. Not only would it prove more complex but it would provide more shocks to the scientific consensus. Working at the University of Montana in 2015, the biologist Toby Spribille was puzzled by the problems encountered in trying to culture lichen in the laboratory. The fact that this could be done had, of course, been essential in providing experimental proof of Schwendener's theory, but it was difficult to accomplish and the results rarely resembled the naturally occurring lichen. The surface layer of the lichen, its cortex, in particular, remained only partially formed. It was one of the spurs to his investigation.

Spribille had an intriguing background. He had been raised in a Montana trailer park and home-schooled in what he described as a 'fundamentalist cult'. Unable to pursue his passion for science and needing desperately to escape from his restrictive upbringing, he took a job in the local forestry service until he had earned enough money to escape from home and seek the education for which he longed. His few savings and lack of qualifications meant that no American university would take him, but learning that some in Europe charged no fees, he made use of his German-speaking family background and secured himself a place at the University of Göttingen. As both an undergraduate and postgraduate he studied lichen, having become familiar with them in his local forests, and in 2011 returned to Montana to join the laboratory of John McCutcheon, a specialist in symbiotic studies.

Knowing the local lichen well, he became interested in the difference between two closely related species of 'horsehair' lichen, *Bryoria fremontii* and *Bryoria tortuosa*. Growing on trees in the mountainous forests of north-west America, they hang in shaggy manes from the higher branches, their tresses as much as ninety centimetres long. *Bryoria fremontii* is dark brown in colour, like cocoa fibre; its relative, more yellowish. It was a surprise, therefore, to find that there was no genetic difference between them. Rerunning his experiments several times, he decided that they were, in fact, a single species, but one with varying environmental forms. Both contained a substance called vulpinic acid and it was the differing quantities of this which led to the difference in form; the more acid, the darker the colour. More surprising still was the fact that the amount of vulpinic acid was also related to the presence of a third organism within each lichen, a species of yeast.

Working with a team of colleagues, Spribille went on to look for similar yeasts in a wide range of other lichen. Using the same techniques they studied fifty-two genera drawn from all six continents, publishing their results in 2016. The yeasts, they concluded, were present in all the lichen, they belonged to a completely new order, and they had been around for at least 200 million years. The lichen was no longer an amalgam of two living things; it was three.

The presence of another organism does not necessarily mean that it has a functioning symbiotic role. It could be a passenger, or even a parasite, but the universal occurrence of these yeasts made this unlikely. Further research soon suggested that they had a role in the production of a protective poison. The toxic quality of some lichens had long been known to native peoples and the wolf lichen had once been used in Scandinavia, stuffed into reindeer carcasses, as a poisonous bait for wolves. Here, it was suggested, the lichen itself was harnessing its poisonous qualities as a deterrent for grazing animals.

With the newly discovered yeast turning out to be ubiquitous, and with lichen having been the subject of so much scrutiny, it might seem surprising that no one had spotted it before. As Spribille himself commented, 'it is remarkable that the Cyphobasidium yeasts have evaded detection, despite decades of molecular and microscopic study'. In fact the yeast was not easily observable under the microscope and it had taken Spribille himself five years to locate it. He had done so using a technique called fluorescent in situ hybridisation, or 'FISH', finding it deeply embedded in the lichen cortex where it spent the entirety of its existence. It was the presence of this third organism that explained Spribille's original conundrum; the reason the lichen were so resistant to culture in the laboratory. An ingredient, it was now clear, had always been missing, and one which helped determine their overall shape. There was a wider significance, too, on which Spribille was clear. The 'fact' that lichen were composed of two species had, he commented, been 'so central to the definition of the lichen symbiosis that it has been codified into lichen nomenclature. This definition had brought order to the field but may also have constrained it'. If the best-known example of symbiosis in science was proving to be more diverse, what else might now be found? Once again the textbooks would need to be rewritten.

* * *

'Are lichens still to be considered as a cosy mixture of plant and fungi', asked Brian Fox in his 1995 Presidential address to the British Lichen Society, adding with a rather English under-statement, 'or is there much more to this plant than meets the eye?' The question was already rhetorical. Work by other researchers had revealed the presence of a much wider array of micro-organisms in the lichen and to some it seemed reasona-ble to wonder whether some of these might not also be part-ners in the symbiosis. Others, like Frank Brightman at Cambridge University, still dismissed such an expansion of the ideas as excessive saying that it was 'unlikely to be very near the truth'. Nonetheless the evidence kept accumulating, with the organisms potentially involved beginning to reach a stag-gering number. In this work the tree lungwort, the same species that Dave and I had pursued through the Wealden woods, was playing a significant part. Two Austrian microbiologists, Martin Grube and Gabriele Berg, had concluded in studying it, that it contained more than 800 species of bacteria and that they bestowed a range of benefits including the provision of nitrogen, the cycling of nutrients, the breaking down of toxins, and the supply of vitamins as well as a range of other chemicals. Elsewhere, researchers in Belfast were finding not just bacteria but a range of other organisms including microfungi, protista and single-celled 'animals' such as amoeba. Further work in Colorado added the microscopic and enigmatic 'wheel animalcules' to the growing list, as well as tardigrades. Now working on the wolf lichen, Spribille himself was establishing the existence of other functioning partners and began to suggest that rather than a set of separate but interacting components, lichen should be regarded as a single 'dynamic system'. They were no longer individuals; they had become a collective.

As with the discovery of symbiosis, what applied to lichen was soon being applied to the rest of the natural world. Every living thing, the microbiologists were now pointing out, was multiple. 'Without symbiosis', wrote Jan Sapp in 'The Symbiotic Self', 'nothing in biology makes sense. A new understanding of life is emerging today, one in which organisms are conceived of as multigenomic entities, comprising many species living together. We are genetic and physiological chimaeras.'

In June 2018, the magazine *British Wildlife* had published an article entitled 'What is a Lichen?' It discussed recent research in the field concluding that a lichen should best be considered as a community. Two years later, James Merryweather, an expert on fungal mycorrhiza, followed it up, only slightly tongue in cheek, with an article entitled, 'What is a Tree?' 'Is a tree just an upside-down lichen?' he asked, perhaps rhetorically. 'Probably more than 50% of its bulk is below ground, a diverse community of symbiotic fungi . . . an integral, inseparable part of any tree, which I contend would make the entire 'community' that is a tree the logical equivalent of a lichen . . . It's just that we typically see it the other way round with half of the tree absent from view.' Nor should we necessarily see the visible 'tree' as the dominant partner. In those updated textbooks we should be 'presenting trees not simplistically as mere individuals, but as modern synthesis of all we know about trees, fungi and other associated organisms in their interactive ecological contexts . . . a tree is not really an individual, it is an ecosystem'.

What does all this say about the human? After all, like both the tree and lichen, there is much more to us than 'meets the eye'. The Human Microbiome Project, conducted in 2021, suggested that there were at least 10,000 species of micro-organisms in the human body. They occupy every part of us from eyelashes to toenails: bacteria, fungi, archaea, viruses, phages, and micro-animals like mites. Though we have been aware of them for some time, it is only recently that we have come to understand their beneficial role. They boost our immunity, ward off pathogens, provide vitamins, help in digestion and promote differentiation of cell tissue. They even affect our behaviour and mood. Without them, in short, we would not be alive. The bacteria alone have been estimated to number around one hundred trillion, outnumbering our own biological cells by a factor that has been estimated at three to one, and sometimes as high as ten to one, making us more bacteria than human. We are, in short, a multitude, a 'we' and not a 'me', and my personal identity crisis is no longer existential but biological as well. Perhaps, as James Merryweather had argued with the tree, it is not necessarily the human in me that should be seen as the dominant partner. The bacteria outnumber me, they evolved with me, they shape me and drive me and some of them will remain when I am gone. Should Bob Gilbert or the bacteria appear on the title page of the book?

* * *

It was in this state of uncertainty that I set out once more to look at lichen. It appealed to me that the organism that had repeatedly raised such challenging questions was not something rare or obscure, to be looked for in distant regions or located only in the laboratory. It was ubiquitous, it was all around us and it was visible everywhere. I had my own mental map of it around my local streets in London's East End. I knew it from the brick walls of the Teviot Estate, a deep rusty orange or granular blue-grey or white with apricot cups. I wondered about its seeming preference for red brick over the stock London yellow. I knew where to find it on tarmac and pavement in the rather anaemically green growth that I knew as 'chewing gum lichen'. It was a resemblance that had sometimes led me to play a game with my children, determining which was lichen and which was gum, though I'm not sure how much it had really entertained them. Its distribution here was something else to think about, for on some surfaces its spreading circles grew densely into each other, while on others it remained completely absent, particularly on the paler, modern, manufactured slabs. On St Leonard's Road, I knew lichen from atop iron railings which supported frondose rosettes of lemon yellow, flecked with paler orange and grey and pitted like the surface of the moon. There was more on wall tops, a dull grey species dotted with black, and on an old wooden rail beside the children's playground patches so bright that they rivalled the buddleias blooming beside the nearby canal. Why was it, I wondered, that the one was admired while the other remained unremarked? I had learnt, too, which were the street trees to scan. The trunk and branches of maple, poplar and narrow-leaved ash were rich with multiple species; the smooth and glossy-barked cherries much less so. The fissured limes attracted a green and granular growth but the peeling bark of the plane could sustain very little. I had even learnt to look out for what Oliver Gilbert called the 'canine zone' at the base of trees where they were most affected by urinating dogs. It could be picked out clearly on the Norway maples along Poplar High Street, a splash zone, so to speak, darkened by the growth of algae and quite clearly demarcated by the lighter growths of lichen in the area above. Dogs, too, were part of the ecology of the lichen.

I did not even have to go as far as the street. After any sort of wind, my garden was littered with twigs and small branches from high in the plane tree. They bore bright patches of the lichen *Xanthoria parietina* which, unusually, also has a plethora of common names. It is the golden lichen, golden shields, yellow scale, sunburst or bronze moss. It was once used in Derbyshire in the traditional well dressing and, under the name of 'crottal', had supplied a small industry on Harris in the dyeing of the local tweed.

All this was interesting enough but I wanted to end my investigations with something special. I came back again to Oliver Gilbert, whose book on lichen was among the first I had read. In its opening chapter he had quoted A.C. Benson. 'Beauty?' he had asked. 'What is it? It is only a trick of old stone and lichen in sunlight'. He had then added his own comment: 'Who could fail to respond to the lichen-covered gods from Stowe Park or the lichen-clad graves of Scottish Kings at Iona Abbey?' And that, I decided, was what I finally had to do, to go to Stowe and Iona and to see these amazing organisms at their recommended best.

* * *

I drove to Stowe with my friend Graham on another sparkling blue spring day, passing through the London suburbs where the planes were in leaf and only the ashes still remained hesitant. Graham had major building works going on in his house and was regaling me with details of cornices and architraves. I only regained full consciousness some time later as we negotiated the repeated roundabouts of Milton Keynes. Milton Keynes was orderly and neat, the old centre of Buckingham was neat, even the surrounding fields were tidy. It reminded him, Graham said, of Switzerland but without the mountains. Such is the size of the Stowe estate that its driveway runs almost all the way from Buckingham, dead straight between rows of trees but switch-backing gently towards its goal; a giant, triumphalist, yellow arch. Beyond the arch is yet another vista and one that is breath-taking in scale; it runs downhill in a broad swathe between flank-ing trees, to a large, ornamental lake, and then up and up again across broad lawns to the house itself, rather severe on the distant ridge. Today the home of Stowe School, it was built by the power-ful Temple-Grenville family in the eighteenth century, and with so many rooms that even Queen Victoria was said to have been bewildered by it. The vast grounds are dotted with follies. Perhaps in a pun on the family name, they include the Temple of Friend-ship, the Temple of Ancient Virtue and the Temple of Concord and Victory, any one of them large enough to happily house several families. We walked past them through the grounds, through the 'Bell Gate' and over the 'Palladian Bridge', in pursuit of the 'Gods of Stowe' finding them eventually in a corner of the Hawkswell Field. The originals had been carved in 1773 by John Michael Rysbrack to represent the Anglo-Saxon gods and as part of a theme that ran through the embellishment of the estate; a celebration of what the nobility saw as an ancient 'English liberty'. With insufficient dedication to the cause they had sold them in 1921 and the ones here today, which Gilbert would also have seen, were replicas.

The gods take their earthly form as a series of statues: classi-cally sculpted figures set on plinths, each adopting a specific pose and bearing their own symbolic attribute. Arranged in a square around a small area of lawn, they are surrounded by a screen of yew in which a goldcrest was feeding as we arrived. They were not as large as I had imagined, and, for deities, not as daunting; in fact they had a rather domesticated feel. They were already heavily worn, even if they were later replicas, and it took us some time to decipher and work out their order. Graham resorted to making tracings of the almost illegible inscriptions and we found our way around the names that had given rise to the days of the week: Sunna, Mona, Tiw, Wotan, Thuner, Friga and Seatern. Thuner was missing from his plinth and I couldn't shake off the impression that he might have popped off somewhere but would soon be returning. Rather than divine permanence, there was a sense of gentle decay about them, their details disappearing, both to erosion and beneath the lichen that now coated them. The gods were clearly not eternal, though the lichen might be.

Across Tiw, they had colonised his feet and his cloak, and his face had been coated completely. The same thing was happening to Wotan. I wondered for a time whether it was almost sinister, as if the lichen were some alien beings, like a Wyndham creation, engulfing these forms and slowly devouring their details. I decided instead it was making them venerable. It was the antidote to their domesticity, and time, after all, did not need the minutia.

In truth, for all of Gilbert's enthusiasm, I had seen more licheniferous structures, old benches in Cornwall, a stone wall in north Wales, but beauty was not just 'old stone and lichen in sunlight', it was in the setting as well, and for some time I sat on the grass amongst the mole hills and dandelions and daisies, a hawk drifting overhead, in the centre of those figures and their lesson in mortality, and it was perfect enough for now.

* * *

Much further away for a Londoner, Iona is a small Hebridean island, just three miles long, and set off the south-west corner of Mull, itself another, if larger, island. It has played a significant part in the life of my family and I was pleased when, a few months later, we were able to set off there again. It takes us the best part of two days to get there and we have always tried to see them as part of the pilgrimage: the train to Glasgow, the walk between stations, another train to Oban, a ferry to Mull and the bus journey that runs over the pass below Ben More and along the Ross of Mull where we look out for red deer, seals and otters. There is a final short ferry ride across the Sound of Mull from which we can see the Abbey standing proud on the sloping east coast below the hill known as Dun I. Surrounded by a little cluster of other buildings, it resembles a mother hen gathering in her brood. It is here because this is the place where Columba settled in 563 and began the spread of Christianity across Scotland and into northern England. It was to become a centre of both political and religious power, for trade and traffic ran, not through the more inhospitable inlands, but, from Iceland and Norway down to Ireland and Man, on the seas that threaded these indented coasts and innumerable islands. Eventually falling into neglect, it was restored in 1938 by unemployed workers from Glasgow, together with trainee ministers from the Church of Scotland, under the leadership of George MacLeod. Here he also founded the now worldwide Iona Community, a radical ecumenical Christian community with its headquarters in Glasgow, but its heart in Iona where it maintains a permanent residential presence at the Abbey. It was with the community that we had come to stay.

* * *

The graveyard is known as the 'Reilig Oran' and sits next to the Abbey, linked to it by the ancient 'Street of the Dead'. Within its low stone wall stands the simple Oran chapel with a wonderful dog-tooth, round-headed doorway and little clumps of wall fern piercing its old stone walls. Around it the graves of ordinary islanders mingle with those of the ancient kings. A survey in 1549 listed fifty-six of them from Scotland and the local kingdom of Dalriada, from Norway and from Ireland. The grass was long when I visited, its flowering heads mixed with the red-tinged sorrel and tall stems of plantain topped with creamy-white anthers. Where the bare rock poked through the thin soil it was coated with mats of stonecrop, ferns and cushions of moss. Atop a mound, the late bluebells were finishing, and hidden here and there in the higher growth were the spreading white flowers of star-of-Bethlehem. Blackbirds flew low, but cluttering loudly, into the neighbouring copse of sycamores, swallows swooped in and out of the open chapel door and from the distant fields I could hear the incessant grate of the corncrake.

Among the traditional upright headstones, the pillars and the needles, many more were laid flat or formed just fractured fragments or small projecting rims of stone like outgrowths from the land. From the older slabs, all trace of inscription had been erased by time and weather and the great and multiple monarchs reduced to an equal anonymity. At Stowe the brighter lichen had contrasted with the pale grey stone, here it was the lichen that were predominantly grey; pale grey, dark grey, dove grey, knife grey, rainy day grey, and they contrasted on many of the stones with a red Mull granite. Some were covered with mats of a flat white growth, dotted with warts, as though they were crusts of a dried and lumpy porridge, while others resembled complex cloud maps. Some covered the whole stone in multiple shades and shapes that intersected in complex curving lines like the divisions in an intricate jigsaw or the sort of line that Paul Klee might take for a walk. Here, too, there were tufted lichen, of the form known as 'fruticose', white in colour or the palest of greens, often in quantity and on one upright stone, packed as thick as a bear's fur. It occurred to me that I had seen such an assemblage before. Here, just a few hundred yards from the sea and regularly raked by the salt-laden winds, the tumbled slabs and stumps and the upright stones were a graveyard equivalent to that grey zone that characterised the lichen banding of the boulder-strewn shore.

I was still in the Reilig Oran as evening came on. The bright glints went out of the sea and the turquoise deepened. Clouds gathered over distant Ben More and the abbey gathered shadow, just as earlier it had gathered its chicks, but across the Sound the rocks of the Ross were bathed in a soft pink light. It seemed an appropriate place to end this journey, here on Iona with its Abbey community and among these ancient stones and the lichen that outlasted inscriptions. 'In the lichen lines on a wall', Leonardo da Vinci had once written, 'an artist should be able to discover a whole landscape.' In fact the lichen had not just revealed a landscape but opened a whole way of seeing. The early botanists had sought to place them in an appropriate taxonomic group but they turned out to belong to both many and none. The lichen was legion and its study had opened our minds to new possibilities; of a process of evolution that was driven as much by co-operation as by competition, of life forms that had multiple points of emergence, of species that merged into each other, of individuals that were composed of multitude, and that we are all of us, in some way, embodied community.

Afterword

For me, there are no answers, only questions, and I am grateful that the questions go on and on.

P.L. Travers

The multiple identity of the lichen, and its implications of community, seemed a fitting place in which to close my 'casebook'. Surveying the lichen in the graveyard of the kings, I had completed the last of my investigations and found a solution of sorts to all of the conundrums I had initially set myself, even if the question had sometimes changed in the process of being answered. If I had set out, however, expecting the neat conclusion of a piece of detective fiction, it was only to discover that the stories did not provide that sort of convenient closure. Instead, every answer seemed to lead to further questions. It was like opening a set of Chinese puzzle boxes, except that every box got bigger rather than smaller, or perhaps like one of those discussions with a child where the answer to every question leads to an insistent, further 'Why?' and the everyday enquiry you started with gets bigger and bigger until you eventually find yourself flailing with God, the Universe and Everything. The files, I found, were not being closed, nor neatly returned to the metaphorical shelf. They were suggesting, instead, a perpetual expansion of possibilities as my initial questions and the answers I found for them led on to bigger places: to re-evaluations of classical taxonomy and of traditional Darwinism, to speculations on the concept of species, and to considerations on whether such a thing as an 'individual' could really be said to exist. They were bringing me face to face with the scale and grandeur of both inner and outer space, and even, through the good graces of the tardigrade, to questions on the way we define life and death. Here, in the end, were confrontations with the incomprehensible. '[We] patiently add fact to fact', Robert Macfarlane has written in his introduction to Nan Shepherd's book *The Living Mountain*, 'but such epistemological bean-counting will only take you so far. No, knowledge is mystery's accomplice rather than its antagonist ... The universe merely refers you onwards.'

And the universe was doing just that. In fact I was increasingly finding that mystery was not just an accomplice but a welcome companion. If it is often seen as a form of darkness, then it is a darkness that is also the imaginative source. It is, said Albert Einstein, 'The most beautiful experience we can have ... the fundamental emotion that stands at the cradle of true art and true science'. This openness to mystery, I was reminded, is a thread that runs through all the world's religions. It is there in Buddhism, in the Sufism of Islam, in the Vedic scriptures of Hinduism and in the Jewish Kabbala. In Christianity it is found in the teachings of Duns Scotus, Julian of Norwich, Margery Kempe and the wonderful Meister Eckhart, and in writings such as the anonymous 'Cloud of Unknowing'. It is a tradition that sits easy with uncertainty and which celebrates an ultimate unknowability, one which understands that the sequences of child-like questions may never have an end. It is beautifully expressed in the long poem that ends the biblical book of Job. Afflicted with multiple losses and great suffering, Job demands answers from God. Instead of answering his questions, God responds with an unanswerable litany of his own:

Who is this that darkens counsel by words without
knowledge? ...
Where were you when I laid the foundation of the earth?
Tell me if you have understanding.

Who determined its measurements – surely you know!
Or who stretched the line upon it?
On what were its bases sunk
or who laid its cornerstone
when the morning stars sang together . . .?
Have the gates of death been revealed to you,
Or have you seen the gates of deep darkness?

For some, however, such a radical uncertainty, such a welcoming acceptance of the unknowable, is too loathed and dreaded and the same world faiths contain another and opposing strand. It was to be found in the fundamentalism with which I was brought up and it sees itself, wherever it occurs, as the sole possessor, and fierce protector, of an ultimate and absolute truth. Its offer is a short-term certainty based on a long-term illusion. Science, too, has been afflicted with a fundamentalism of its own, one which still persists in some areas. We have come from a more self-confident world, one in which knowledge, logical, rational, testable knowledge, was seen as continually driving back darkness. It was a self-confidence fuelled by the Enlightenment and by a professionalisation and privatisation of knowledge, lacking what T.S. Eliot called the 'wisdom of humility'. Despite its many advances, Newtonian physics, together with Descartian materialism, helped fashion a world where facts, though multitudinous, were seen as finite, and in which we would one day, however distant, uncover them all. The universe was a vast but predictable machine that we could tinker with and perfect. It was a view that drove poets to despair, most famously, John Keats in his long poem 'Lamia':

Philosophy will clip an Angel's wings,
Conquer all mysteries by rule and line,
Empty the haunted air, and gnomed mine,
Unweave a rainbow . . .

Even Albert Einstein was once led to remark that 'the process of scientific discovery is a continual flight from wonder'. But it is not. It is the very opposite; it is a deep and continuous journey in which the universe is revealed as increasingly awesome. Science, today, sits more comfortably with uncertainty; with chaos theory, relativity, the uncertainty principle, quantum physics and the unidentified majority of matter, with the fifth force of nature, with planetary bodies that number in their trillions and with the unimaginable vastness of both space and time.

In the openness of uncertainty, mystery is the place where science and faith can meet. Keats would be welcome here, too, for it is from mystery and from the same power of the imagination that the arts and sciences, so roughly and so wrongly divided, both spring. Perhaps I am trying too hard to have my celestial cake and eat it too. Nonetheless, as I sought out the solutions to the cases in this book, the missing fragrance of the musk, the ancient associations of the yew, the unexplained appearances of star jelly, the secret identity of lichen, the story of the Underground mosquito and the uncertain origins of the super-tough tardigrade, and as I came across the often amazing answers to my questions, those answers continually led to places of greater awe and wonder. As the nature writer D.W. Gillingham wrote in the 1930s: 'Science has opened our eyes for good or evil . . . rather than explaining all things to us, it has made the familiar more magical still. Science cries "Look!"' The task, I had found, was not to dispel the mystery. It was to enter it more deeply.

Acknowledgements

I sometimes suspect that the only people who read acknowledgements are those who are looking to see if they're mentioned, which is fine except that I would really like the chance to pay wider tribute to the institutions in this country which still provide their time and services to the public for free, and which were indispensable in the writing of this book. Among those I made particular use of were the British Library, the library of the Wellcome Collection, the library of the London Natural History Museum (as well as its Angela Marmont Centre), the library and herbarium of the Royal Botanic Gardens, Kew, and the library and archives of the Royal Horticultural Society. The London Transport Museum at Covent Garden is not free, but it was still an invaluable resource. May they all, and their helpful staff, long continue.

As well as these organisations and institutions, I am grateful for the willingness of individual experts in many fields to give me their time, share their knowledge and answer my many questions. Among them I should particularly mention Dr Gothamie Weerakoon of the Natural History Museum and Professor Ian Kinchin of the University of Surrey. Florence Matteson helped me with information on genetics, Amanda Hopkinson undertook translations for me, Heather Walton discussed the role of mystery and the woodturner Darren Breeze taught me about the properties of yew wood – as well as using it to turn me the miniature tree that stands on my desk as I write. Chris Carter freely shared his knowledge of algae, as well as his beautiful photographs, and pointed me in the direction of several highly useful sources. Jan Sellars, then Clerk of Wanstead Quaker Meeting, located Gulielma's Quaker testament for me and I am grateful to her and other members of the

Meeting who provided me with information on the life of this remarkable woman.

My various expeditions about the country, in search of face-to-face encounters with my subject matter, were vital components of the book. I would like to thank all those who were so willing to accompany me. My Fortean friends from Edinburgh, Ann Jones and Stuart Robins, drove me about Scotland in search of ancient yews and other remarkable phenomena. Graham Scrivener accompanied me to Stowe, while at Chigwell School, Nicole Benjamin, Derek Wyatt-Barrett and Chris Lord assisted in the ultimately successful search for a tardigrade. My friend Brother Sam SSF discussed many ideas for the book with me as we walked through the Essex countryside, and Nick Danby, and the staff of the Seacourt Real Tennis Centre, could not have been more helpful when I turned up there unannounced. Julian May, the producer of many excellent programmes for Radio 4, accompanied me in pursuit of ancient yews in Surrey while, closer to home, Al Dix helped me to explore the microscopic world – and assisted in the depletion of my whisky collection. Since I make particular mention in the book of his failure to sit his Biology A level, I feel it only fair to mention here that he is now the holder of an Honorary Doctorate. As ever, the indefatigable Dave Bangs rose to any challenge that I cast before him and our adventures in search of plants, lichen, mosses and slime moulds were always accompanied by long discussions on life, love and politics – especially the question of public access to the countryside.

These memorable outings aside, writing is a solitary occupation and I am not sure I am constitutionally cut out for it. I am particularly grateful therefore to the friends who provided support and encouragement and kept me going along the way. Heather Walton helped me with theological discussions and provided me with my opening quote; Sarah Cleave provided information and sources on 'astromyxin'; Maggie Lippiett introduced me to the poetry of Andrew Young; Adrian Newman, Francis Straw, Al Dix, Graham Scrivener, Peter Lippiett, the members of the Mehitabel writers' group and my always supportive agent, James Macdonald Lockhart, all read and commented on sections of the text whilst I promised Fraser Dyer a mention for unhesitatingly handing over his best pen on the occasion when mine ran out. In a small room in Iona Abbey, Meg Wroe, Martin Wroe and Molly Boot patiently listened to a reading of what became the introduction and conclusion and helped me greatly with their comments. My thanks go to my editors at Sceptre, Juliet Brooke, Jo Dingley and Holly Knox, who helped me shape the narrative and curb the more digressive of my digressions, and I am particularly grateful to John Swindells who read and corrected every chapter, gave freely of his botanical knowledge and introduced me to the musk storksbill. As with my last book, Rod Harper and Gordon Willis were my essential back-up team, not just reading each chapter as it was completed but keeping me going with their positivity and encouragement.

It is almost traditional to end acknowledgements with a mention of your family, but in this case I am genuinely both happy and proud that *The Missing Musk* was virtually a family production. My sister Gill read sections of the text and took me on expeditions around Hayling Island, as well as buying me dinner in The Yew Tree pub. Most of all she told me for years that this was the book she wanted me to write and harried me to get on with it. Each of my three sons also played a part. Matthew was positive and encouraging, while Joel (and Rosie) read early drafts. I believe it may even have played a role in their courtship. Thomas provided the excellent illustrations, even wrestling with the conundrum of how to portray a lump of jelly. My wife Jane, as ever, provided her support and encouragement, particularly during my lowest moments. Apparently Virginia Woolf used to spend each morning sitting in her bath reading aloud whatever she had written the previous day. I was always envious of that and grateful that we were able to invent our own alternative. Every Saturday morning, which was Jane's only day off work, we would sit together in bed while I read her some, or all, of the previous week's writing. They were some of the happiest moments in the making of this book.

My love and thanks go to you all.

Select Bibliography

CHAPTER 1: THE MISSING MUSK

Banks, G., *The Plymouth and Devonport Flora* (privately published, 1830–32)

Brown, Jane, *The Pursuit of Paradise* (HarperCollins, 1999)

Burton, Rodney, *Flora of the London Area* (London Natural History Society, 1983)

Burton, Rodney, 'The Changing Status of Musk Stork's-bill' (*London Naturalist*, vol. 84, 2005)

Clapham, A.R., Tutin, T.G., Warburg, E.F., *Flora of the British Isles* (Cambridge University Press, 1962)

Davies, John, *Douglas of the Forests: The North American Journals of David Douglas* (Paul Harris Publishing, 1979)

Douglas, David, *Journal kept by David Douglas during his travels in North America 1832–1837* (W. Wesley and Son, 1914)

Genders, Roy, *The Scented Wild Flowers of Britain* (Collins, 1971)

Genders, Roy, *Scented Flora of the World* (Robert Hale, 1977)

Gerard, John, *The Herball or General Historie of Plantes* (John Norton, 1597)

Greenwell, Jean, 'Kaluakauka Revisited: The Death of David Douglas in Hawaii' (*Hawaiian Journal of History*, vol. 22, 1988)

Harvey, John, *Early Gardening Catalogues* (Phillimore, 1972)

Hope, Frances Jane, *Notes and Thoughts on Gardens and Woodlands* (Macmillan, 1881)

Keble Martin, Rev. W., *The Concise British Flora* (Ebury Press, 1965)

Matthew, H. and Harrison, B. (eds), *Dictionary of National Biography* (Oxford University Press, 2004)

Stace, Arthur J., 'Plant Odors' (*Botanical Magazine*, vol. 12, no. 11, November 1887)

Stuart, D. and Sutherland, J., *Plants from the Past* (Viking, 1987)

Wilson, E.H., *A Naturalist in Western China* (Methuen, 1913)

CHAPTER 2: DEATH AND LIFE IN THE CHURCHYARD

Ablett, W.H., *English Trees and Tree Planting* (Smith, Elder and Co., 1880)

Bevan-Jones, Robert, *The Ancient Yew: A History of Taxus Baccata* (Windgather Press, 2002)

Buczacki, Stefan, *Earth to Earth: A Natural History of Churchyards* (Unicorn, 2018)

Chetan, Anand and Brueton, Diana, *The Sacred Yew* (Penguin Arcana, 1994)

Christie, Agatha, *A Pocketful of Rye* (HarperCollins, 1953)

Cornish, Vaughan, *The Churchyard Yew and Immortality* (Frederick Muller, 1946)

Dearmer, P., Vaughan Williams, R., Shaw, M., *The Oxford Book of Carols* (Oxford University Press, 1928)

Forsyth, A.A., *British Poisonous Plants* (Ministry of Agriculture, Fisheries and Food, Bulletin no. 161, HMSO, 1968)

Fortingall Church, *Fortingall: Kirk and Village* (Church guide, 2015)

Grieve, Maude, *A Modern Herbal* (Jonathan Cape, 1931)

Grigson, Geoffrey, *The Englishman's Flora* (Phoenix House, 1955)

Hageneder, Fred, *Yew* (Reaktion Books, 2013)

Hall, Tony, *The Immortal Yew* (Kew Publishing, 2018)

Hutton, Ronald, 'Under the Spell of the Druids' (*History Today*, April 2009)

Johns, Rev. C.A., *The Forest Trees of Britain* (SPCK, 1849)

Kite, G.C. et al., 'Generic detection of basic taxoids in wood of European yew (*Taxus baccata*) by liquid chromatography-ion trap mass spectrometry' (*Journal of Chromatography*, February 2013)

Lowe, John, *The Yew Trees of Great Britain and Ireland* (Macmillan and Co., 1897)

Ness, Patrick, *A Monster Calls* (Walker Books, 2011)

O'Curry, E., *Manners and Customs of the Ancient Irish* (Williams and Norgate, 1873)

Pakenham, Thomas, *Meetings with Remarkable Trees* (Weidenfeld and Nicolson, 1996)

Strutt, Jacob George, *Sylva Britannica* (Longman, Rees, Orme, Brown and Green, 1830)

Vickery, Roy, *Vickery's Folk Flora* (Weidenfeld and Nicolson, 2019)

CHAPTER 3: STAR FALL

Barker, Juliette, *Agincourt: The King, the Campaign, the Battle* (Abacus, 2005)

Beer, Amy-Jane, 'Wild Story' (*British Wildlife*, vol. 40, no. 4, April 2019)

Belcher, Hilary and Swale, Eric, 'Catch a Falling Star' (*Folklore*, vol. 95ii, 1984)

Bronowski, Jacob, *The Ascent of Man* (Little, Brown and Co., 1973)

Bychkov, V.L., 'The Physical Nature of Gelatinous Meteors: "pwdre ser" or "star jelly"' (*International Journal of Meteorology*, vol. 30, October 2005)

Bychkov, V.L. et al. (eds), *The Atmosphere and Ionosphere: Elementary Processes, Monitoring and Ball Lightning* (Springer, 2014)

Fort, Charles, *The Book of the Damned* (Boni and Liveright, 1919)

Hagman, Larry (director), *Beware! The Blob* (Jack H. Harris Enterprises, 1972)

Herzog, Werner and Oppenheimer, Clive (directors), *Fireball: Visitors from Darker Worlds* (Sandbox Films, 2020)

Keller, Verner and Neil, William (trans.), *The Bible as History* (Hodder and Stoughton, 1965)

Lister, Arthur and Lister, Gulielma, *A Monograph of the Mycetozoa: a catalogue of the species in the British Museum* (British Museum, 1935)

Morton, Andrew, *Trees of the Celtic Saints: The Ancient Yews of Wales* (Gwasg Carreg Gwalch, 2009)

Morton, John, *The Natural History of Northamptonshire* (Knaplock, 1712)

Orinski, G.R., Cockerell, C.S., Pontefract, A. and Sapers, H.M., 'The Role of Meteorite Impacts in the Origin of Life' (*Astrobiology*, vol. 20, no. 9, September 2020)

Pitt, Malcolm, 'Etymology of the Genus Name Nostoc' (*International Journal of Systematic Bacteriology*, vol. 47, no. 2, April 1997)

Ramsbottom, J., *Mushrooms and Toadstools* (Collins, 1953)

Russell, Chuck (director), *The Blob* (Palisades California, 1988)

Scott, Walter, *The Talisman* (Archibald Constable, 1825)

Yeaworth, Irvin (director), *The Blob* (Fairview Productions, 1958)

Young, Andrew, *Selected Poems* (Carcanet Press, 1998)

CHAPTER 4: THE UNDERGROUND MOSQUITO

Ackroyd, Peter, *London Under* (Vintage, 2012)

Bankoff, Greg, 'Malaria, Water Management and Identity in the English Lowlands' (*Environmental History*, vol. 23, no. 3, July 2018)

Bates, Claire, 'Would it be wrong to eradicate mosquitoes? (*BBC News Magazine*, 28 January 2016)

Baylis, Matthew, 'Potential impacts of climate change on emerging vector-borne and other infections in the UK' (*Environmental Health*, vol. 16, supplement 1, 2017)

Byrne, Katharine and Nichols, Richard, '*Culex pipiens* in London Underground tunnels: differentiation between surface and subterranean populations' (*Heredity*, vol. 82, January 1999)

Chinery, Michael, *Collins Guide to the Insects of Britain and Western Europe* (Collins, 1986)

Dobson, Mary J., 'Marsh Fever – the geography of malaria in England' (*Journal of Historical Geography*, vol. 6, no. 4, 1980)

Dobson, Mary J., 'When Malaria was an English Disease' (*Geographical Magazine*, February 1982)

Dobson, Mary J., 'Malaria in England: a geographical and historical perspective' (*Parassitologia*, vol. 36, nos. 1–2, 1994)

Eastman Classroom Films, *Life History of the Mosquito Aedes Aegyptus* (Medical School, University of Rochester, 1928)

Hayling Island U3A, *Hayling Island: The Years of Change 1919–1946* (Hayling Island University of the Third Age)

HM Government *National Contingency Plan for Invasive Mosquitoes: Detection of Incursion* (HM Government, May 2020)

Khun, K.G. et al., 'Malaria in Britain: Past, present and future' (*Proceedings of the National Academy of Sciences of the United States of America*, vol. 100, no. 17, August 2013)

MacArthur, Sir W., 'A Brief History of English Malaria' (*Post Graduate Medical Journal*, vol. 22, 1946)

Marren, Peter and Mabey, Richard, *Bugs Britannica* (Chatto and Windus, 2010)

Marshall, J.F., *The British Mosquitoes* (British Museum, 1938)

Mourier, Henri and Winding, Ove, *Collins Guide to Wild Life in House and Home* (Collins, 1975)

Reiter, Paul, 'From Shakespeare to Defoe: Malaria in England in the Little Ice Age' (*Emerging Infectious Diseases*, vol. 6, no. 1, February 2000)

Reznick, David, *The 'Origin' Then and Now* (Princeton University Press, 2011)

Schilthuizen, Menno, *Darwin Comes to Town: How the Urban Jungle Drives Evolution* (Quercus, 2018)

Sherwood, John, *Creeping Jenny* (Macmillan, 1993)

Shute, P.G., 'Culex Molestus' (*Transactions of the Royal Entomological Society*, vol. 102, 1951)

Snow, Keith and Snow, Susan, 'John Frederick Marshall and "The British Mosquitoes"' (*Bulletin of the British Museum (Natural History)*, vol. 19, no. 1, June 1991)

Spielman, Andrew and D'Antonio, Michael, *Mosquito: The Story of Man's Deadliest Foe* (Faber and Faber, 2001)

Thirsk, Joan and Finberg, H.P.R., *The Agrarian History of England and Wales: Volume 4, 1500–1640* (Cambridge University Press, 1967)

Winegard, Timothy C., *The Mosquito: A Human History of Our Deadliest Predator* (Text Publishing, 2019)

CHAPTER 5: THE INTERGALACTIC BEAR

Atherton, I., Bosanquet, S. and Lawley M., *Mosses and Liverworts of Britain and Ireland: A Field Guide* (British Bryological Society, 2010)

Baertierchen/Tardigrade website: tardigrades.com

Bartels, Paul, *Water Bears of the Smokies and Beyond* (YouTube talk, Science@Sugarlands series, 20 August 2021)

Cox, Brian, *Are we thinking about alien life all wrong?* (BBC Ideas, 29 November 2021)

Crick, Francis, *Life Itself: Its Origin and Nature* (Simon and Schuster, 1981)

Fleischfresser, S., 'A Brief History of Panspermia' (*Cosmos Magazine*, 23 April 2018)

Fleischfresser, S., 'Viruses, ET and the Octopus from Space: the return of panspermia' (*Cosmos Magazine*, 26 April 2018)

Fleming, Anthony, 'Bears in Britain' (*British Wildlife*, vol. 19, no.6, August 2008)

Gross, V. et al., 'Miniaturization of Tardigrades: Morphological and genomic perspectives' (*Arthropod Development and Growth*, vol. 48, January 2019)

Hoyle, F., *The Intelligent Universe* (Michael Joseph, 1983)

Hoyle, F. and Wickramasinghe C., *Evolution from Space* (J.M. Dent, 1981)

Kimmerer, Robin Wall, *Gathering Moss* (Penguin, 2021)

Kinchin, Ian M., *The Biology of Tardigrades* (Portland Press, 1994)

Marren, Peter and Mabey, Richard, *Bugs Britannica* (Chatto and Windus, 2010)

May, Andrew, *Astrobiology: The Search for Life Elsewhere in the Universe* (Icon, 2019)

Miller, W., 'Tardigrades' (*American Scientist*, vol. 99, no. 5, Sept–Oct 2011)

Morgan, Clive I. and King, P.E., *British Tardigrades* (Linnaean Society/Academic Press, 1976)

Pierres, Marianne de, *Dark Space* (Orbit, 2007)

Plaskitt, F.J.W., *Microscopic Fresh Water Life* (Chapman and Hall, 1926)

Rampelotto, P.H., 'Resistance of Microorganisms to Extreme Environmental Conditions and Its Contribution to Astrobiology' (*Sustainability*, vol. 2, no.6, June 2010)

Ruvkun, Gary, *What is true for e-coli on earth will be true for life on Proxima Centauri B* (YouTube Breakthrough Discussion, UC Berkeley, April 2019)

Schill, R.O. (ed.), *The Biology of Tardigrades* (Springer, 2018)

Shklovskii, I.S. and Sagan, C., *Intelligent Life in the Universe* (Picador, 1966)

Sloan, D., Batista, R.A., Loeb, A., 'The Resilience of Life to Astrophysical Events' (*Scientific Reports*, July 2017)

Steele, Edward et al., 'Causes of the Cambrian Explosion – Terrestrial or Cosmic?' (*Progress in Biophysics and Molecular Biology*, vol. 136, August 2018)

Traspas, A. and Burchell, M.J., 'Tardigrade Survival Limits in High-speed Impacts – Implications for Panspermia' (*Astrobiology*, vol. 21, no. 7, July 2021)

Vetter, Joachim, 'The "Little Bears" that Evolutionary Theory Can't Bear' (*Creation*, vol. 12, no. 2, March 1990)

Wickramasinghe, J., Wickramasinghe, C.W. and Napier, W., 'Comets and the Origin of Life' (*World Scientific*, 2010)

CHAPTER 6: THE I IN LICHEN

Ainsworth, G.C., *Introduction to the History of Mycology* (Cambridge University Press, 1976)

British Mycological Society, *Beatrix Potter: Pioneering Scientist or Passionate Amateur?* (Website posting, 20 June 2017)

Dobson, Frank S., *Lichens: An Illustrated Guide to the British and Irish Species* (British Lichen Society/Richmond, 2018)

Gilbert, Oliver, *Lichens* (HarperCollins, 2000)

Gilbert, S.F. et al., 'A Symbiotic View of Life: We have never been individuals' (*Quarterly Review of Biology*, vol. 87, no. 4, December 2012)

Hawksworth, D.L. and Seaward, M.R.D., *Lichenology in the British Isles 1568–1975* (Richmond, 1977)

Honegger, R, 'The Lichen Symbiosis – What is so spectacular about it?' (*The Lichenologist*, vol. 30, no.3, 1998)

Honegger, R., 'Simon Schwendener and the dual hypothesis of lichen' (*The Bryologist*, vol. 103, no. 2, Summer 2000)

Laundon, Jack R., *Lichen* (Shire Natural History, 1986)

Margulis, L. and Kratz, R.F. (eds), *Symbiosis as a Source of Evolutionary Innovation: Speciation and Morphogenesis* (MIT Press, 1991)

Merryweather, James, 'What is a Tree? Changing Minds about Symbiosis' (*British Wildlife*, vol. 31, no. 4, April 2020)

Mitchell, M.E., 'Such a Strange Theory: Anglophone attitudes to the discovery that lichens are composite organisms, 1871–1890' (*Huntia*, vol. 11, no. 2, 2002)

Mitchell M.E., 'De Bary's Legacy: The emergence of different perspectives on lichen symbiosis' (*Huntia*, vol. 15, no. 1, 2014)

Nicholson, B.E. and Brightman, F.H., *The Oxford Book of Flowerless Plants* (Oxford University Press, 2020)

Sapp, Jan, *Evolution by Association* (Oxford University Press, 2009)

Sapp, Jan, *The New Foundations of Evolution: On the Tree of Life* (Oxford University Press, 2009)

Sapp, Jan, 'The Symbiotic Self' (*Evolutionary Biology*, no. 43, March 2016)

Sheldrake, Rupert, *Entangled Life* (Bodley Head, 2020)

Smith, Annie Lorrain, *Lichens* (Cambridge University Press, 1921)

Spribille, T. et al., 'Basidiomycete yeasts in the cortex of ascomycetes macrolichens' (*Science Journal*, vol. 353, no. 6298, July 2011)

Thomas, Lewis, *The Lives of a Cell: Notes of a Biology Watcher* (Viking, 1974)

Wilkinson, David M., 'What Is a Lichen? Changing Ideas on Lichen Symbiosis' (*British Wildlife*, vol. 29, no. 5, June 2018)

Wotton, Roger S., 'What was Manna?' (*Opticon*, 1826, issue 9, Autumn 2010)

Yong, Ed, 'How a Guy from a Montana Trailer Park Overturned 100 Years of Biology' (*The Atlantic*, July 2016)

Yong, Ed, *I Contain Multitudes* (Vintage, 2017)